Strategic Cost Fundamentals

for Designers, Engineers, Technologists, Estimators, Project Managers, and Financial Analysts

Synthesis Lectures on Engineering

Each book in the series is written by a well known expert in the field. Most titles cover subjects such as professional development, education, and study skills, as well as basic introductory undergraduate material and other topics appropriate for a broader and less technical audience. In addition, the series includes several titles written on very specific topics not covered elsewhere in the Synthesis Digital Library.

Strategic Cost Fundamentals: for Designers, Engineers, Technologists, Estimators, Project Managers, and Financial Analysts
Robert C. Creese
2018

Empowering Professional Teaching in Engineering: Sustaining the Scholarship of Teaching
John Heywood
2018

The Human Side of Engineering
John Heywood
2017

Geometric Programming for Design Equation Development and Cost/Profit Optimizaton, Third Edition
Robert C. Creese
2016

Engineering Principles in Everyday Life for Non-Engineers
Saeed Benjamin Niku
2016

A, B, See... in 3D: A Workbook to Improve 3-D Visualization Skills
Dan G. Dimitriu
2015

The Captains of Energy: Systems Dynamics from an Energy Perspective
Vincent C. Prantil and Timothy Decker
2015

Geometric Programming for Design and Cost Optimization (with illustrative case study problems and solutions), Second Edition
Robert C. Creese
2010

Survive and Thrive: A Guide for Untenured Faculty
Wendy C. Crone
2010

Geometric Programming for Design and Cost Optimization (with Illustrative Case Study Problems and Solutions)
Robert C. Creese
2009

Style and Ethics of Communication in Science and Engineering
Jay D. Humphrey and Jeffrey W. Holmes
2008

Introduction to Engineering: A Starter's Guide with Hands-On Analog Multimedia Explorations
Lina J. Karam and Naji Mounsef
2008

Introduction to Engineering: A Starter's Guide with Hands-On Digital Multimedia and Robotics Explorations
Lina J. Karam and Naji Mounsef
2008

CAD/CAM of Sculptured Surfaces on Multi-Axis NC Machine: The DG/K-Based Approach
Stephen P. Radzevich
2008

Tensor Properties of Solids, Part Two: Transport Properties of Solids
Richard F. Tinder
2007

Tensor Properties of Solids, Part One: Equilibrium Tensor Properties of Solids
Richard F. Tinder
2007

Essentials of Applied Mathematics for Scientists and Engineers
Robert G. Watts
2007

Project Management for Engineering Design
Charles Lessard and Joseph Lessard
2007

Relativistic Flight Mechanics and Space Travel
Richard F. Tinder
2006

Strategic Cost Fundamentals: for Designers, Engineers, Technologists, Estimators, Project Managers, and Financial Analysts
Robert C. Creese

ISBN: 978-3-031-79393-6   paperback
ISBN: 978-3-031-79394-3   ebook
ISBN: 978-3-031-79395-0   hardcover

DOI 10.1007/978-3-031-79394-3

A Publication in the Springer series
SYNTHESIS LECTURES ON ENGINEERING

Series ISSN
Print 1939-5221   Electronic 1939-523X

Strategic Cost Fundamentals

for Designers, Engineers, Technologists, Estimators, Project Managers, and Financial Analysts

Robert C. Creese

SYNTHESIS LECTURES ON ENGINEERING #32

ABSTRACT

This book is designed to introduce designers, engineers, technologists, estimators, project managers, and financial analysts as well as students in engineering and business to strategic cost tools for project cost evaluations. The three main sections are as follows. (1) Cost Relationships, Financial Statements, and Performance Measures—This section describes the relationships between cash flows and profits; the relationships between financial statements and the Purcell Diagram; and the issues of cost estimating, time-based breakeven analysis and time-based earned schedule. (2) Tools for Economic Evaluations—This section considers the basic mathematical relations used behind the economic equations and factors; discrete and continuous interest; depreciation terms and methods; and the Present Value of Principal Approach for evaluating loans. (3) Methods for Project Evaluation and Risk Analysis—This section considers payback periods, present worth analysis, return on investment, internal rate of return, benefit/cost ratios and positive-negative project balances; risk techniques of sensitivity analysis, optimistic-pessimistic analysis, discrete probability examples, and continuous probability models using the normal and triangular distributions.

KEYWORDS

risk analysis, project evaluation, loans, Purcell diagram, engineering economic expressions, breakeven analysis, cost estimating and profit calculations, depreciation methods, earned value management

Contents

PART I

Cost Relationships, Financial Statements, and Performance Measures

CHAPTER 1

Fundamental Terms and Concepts

1.1 INTRODUCTION

Strategic cost management includes economic analysis (engineering economics) as well as cost estimating (cost engineering), project management, and financial analysis. Cash flows and profits are the two major strategic cost measures for evaluating the success of an enterprise and taxes, loans, and depreciation have major influences upon the cash flows and profits. Positive cash flows are necessary for the success of an enterprise and is similar to the need of water to keep a plant alive. Positive profits are the rewards of the enterprise and are similar to the fruit of the plant. If there are no positive cash flows, the enterprise will fail similar to a plant without water and die. If the enterprise does not produce positive profits, the enterprise fails as the investors will not support it and plants that do not produce will also be left to die.

Equipment is needed to produce products and the cost of the equipment must be recovered and that is done through the concept of depreciation. Depreciation amounts are spread over the predicted economic life of the investment and different approaches for determining the depreciation amounts can be used. Equipment purchases often require loans to assist in the purchase of equipment and the loan interest is a depreciable expense. Taxes are assessed in a variety of manners, such as on property values or on the amount of profits earned, and are used to provide services for the community where the facility is located. Taxes and depreciation are considered as expenses similar to the raw materials, labor, energy, and other items utilized to produce the products that are sold to produce the profits and positive cash flows.

1.2 BASIC RELATIONSHIPS BETWEEN CASH FLOWS, PROFITS, DEPRECIATION, AND TAXES

Cash flows represent the net monetary units, such as Dollars, Pounds, Euros, Yen, Won, Bitcoins, Pesos, or other currency flowing into and out of a business financial venture and it is desired to have a positive net cash flow. It represents the funds available for expenses and business enterprises must have an adequacy of funds to pay for their expenses. Companies with negative net cash flows will not survive very long, whereas companies with negative profits may survive for several years if they have positive cash flows.

Net Cash Flow is the difference between total cash receipts (inflows) and total cash disbursements (outflows) for a given period of time such as a month, quarter, or year. They represent the funds available to pay for expenses and savings for major future investments rather than borrowing funds.

Profits represent the net revenues minus the expenses and companies must be profitable to survive in the long-term. Short-term periods of losses often occur during the start-up of businesses or during periods of economic recession, but long-term losses are not sustainable for an enterprise. Net Profits are also referred to as Earnings and thus Net Profits per Share is the same as Earnings per Share.

The following relations are utilizing the basic expressions without adjustments. More items can be considered, but these are the primary relationships.

$$\text{Gross Profits} = \text{Revenues} - \text{Costs} - \text{Depreciation} \tag{1.1}$$

$$\text{Net Profits} = \text{Gross Profits} - \text{Taxes} = \text{Gross Profits} \times (1 - \text{Tax Rate}) \tag{1.2}$$

$$\text{Taxes} = \text{Tax Rate} \times \text{Gross Profits} \tag{1.3}$$

$$\text{Net Profits} = (1 - \text{Tax Rate}) \times (\text{Revenues} - \text{Costs} - \text{Depreciation}) \tag{1.4}$$

$$\text{Cash Flows} = \text{Revenues} - \text{Costs} - \text{Taxes} \tag{1.5}$$

$$\text{Cash Flows} = \text{Revenues} - \text{Costs} - \text{Tax Rate}(\text{Revenues} - \text{Costs} - \text{Depreciation}) \tag{1.6}$$

$$\text{Cash Flows} = (1 - \text{Tax Rate}) \times (\text{Revenues} - \text{Costs} - \text{Depreciation}) + \text{Depreciation} \tag{1.7}$$

or

$$\text{Cash Flows} = \text{Net Profits} + \text{Depreciation} \tag{1.8}$$

this can also be written as:

$$\text{Net Profits} = \text{Cash Flows} - \text{Depreciation} \tag{1.9}$$

Note that the tax rate is expressed as a decimal in the formulas, thus a 10% tax rate is expressed in decimal form as 0.10. It can be noted from Equations (1.8) and (1.9) that depreciation has a positive effect on cash flows and a negative effect upon net profits. A decision must be made to select either cash flows or net profits as the primary objective of the corporation and thus one must also focus on the depreciation methods typically used.

There are two major depreciation methods utilized which are the straight line method which gives equal amounts of depreciation per year over the life of the investment or the Modified Accelerated Cost Recovery System (MACRS)—referred to as *accelerated depreciation method* in this chapter) which gives higher depreciation amounts in the early years of the investment life and lower amounts in the later years of the investment. There is an apparent dilemma as accelerated depreciation would initially give higher cash flows and lower net profits than straight line depreciation and that would not please stockholders and investors. So the business community has decided to use both methods; they use accelerated depreciation methods to determine the taxes they report to the government and use the straight line depreciation method to report profits to the stockholders and investors. Thus, the reported profits to stockholders and investors are not the actual profits, but are inflated by the differences in the depreciation methods used. The difference in the depreciation methods adjusted by the tax rate is reported as deferred taxes in the report to the stockholders, but the difference in the net profits is usually not presented in the stockholders' report. The purpose of the accelerated depreciation method was to encourage companies to make investments to modernize and improve their production processes. Since accelerated depreciation improved the cash flows and a straight line was used to report profits to the stockholders, the decision was made to focus upon cash flows rather than profits in evaluating projects and/or investment alternatives.

A second advantage of the focus on cash flows is that in many cases only cost data is available and the prices for the selling of the products is not known, so the alternatives being investigated can be compared on a minimum cost basis rather than a maximum profit basis, but care must be taken in this type of analysis.

1.3   CASH FLOW AND PROFIT EXAMPLE

An example illustration will be presented to show the differences in cash flows and profits using straight line depreciation and accelerated depreciation methods. The data presented in Table 1.1 is used in the formulas presented.

Table 1.1: Cash flow and depreciation data for example problem

Item	Amount ($)	Amount (%)
Revenue	1,000,000	
Costs	775,000	
Tax Rate (%)		25
Depreciable Investment	250,000	
Depreciation—Straight Line	25,000	10
Depreciation—Accelerated (MACRS)	50,000	20

The first analysis will be using the accelerated depreciation method, which is what it will be using to report to the government, and it will be assumed that the company has 10,000 shares of stock.

$$\text{Gross Profits} = \text{Revenues} - \text{Costs} - \text{Depreciation} \tag{1.1}$$
$$\text{Gross Profits} = \$1,000,000 - \$775,000 - \$50,000 = \$175,000$$

$$\text{Taxes} = \text{Tax Rate} \times \text{Gross Profits} \tag{1.3}$$
$$\text{Taxes} = 0.25 \times \$175,000 = \$43,750$$

$$\text{Net Profits} = \text{Gross Profits} - \text{Taxes} = \text{Gross Profits} \times (1 - \text{Tax Rate}) \tag{1.2}$$
$$\text{Net Profits} = \$175,000 \times (1 - 0.25) = \$131,250$$

The earnings per share would be:
Earnings/share = $131,250/10,000 shares = 13.125 $/share.

$$\text{Cash Flows} = \text{Revenues} - \text{Costs} - \text{Taxes} \tag{1.5}$$
$$\text{Cash Flows} = \$1,000,000 - \$775,000 - \$43,750 = \$181,250$$

or

$$\text{Cash Flows} = \text{Net Profits} + \text{Depreciation} \tag{1.9}$$
$$\text{Cash Flows} = \$131,250 + \$50,000 = \$181,250$$

The second analysis will be using the straight line depreciation method, which will be reported to its stockholders, and it will also be assumed that the company has 10,000 shares of stock.

$$\text{Gross Profits} = \text{Revenues} - \text{Costs} - \text{Depreciation} \tag{1.1}$$
$$\text{Gross Profits} = \$1,000,000 - \$775,000 - \$25,000 = \$200,000$$

$$\text{Taxes} = \text{Tax Rate} \times \text{Gross Profits} \tag{1.3}$$
$$\text{Taxes} = 0.25 \times \$200,000 = \$50,000$$

$$\text{Net Profits} = \text{Gross Profits} - \text{Taxes} = \text{Gross Profits} \times (1 - \text{Tax Rate}) \tag{1.2}$$
$$\text{Net Profits} = \$200,000 \times (1 - 0.25) = \$150,000 \text{ instead of } 131,250$$

The earnings per share would be:
Earnings/share (EPS) = $150,000/10,000 shares = 15.00 $/share instead of 13.125 $/share.

The difference in (EPS) = 15.00 \$/share − 13.125 \$/share = 1.875 \$/share

$$\text{Cash Flows} = \text{Revenues} - \text{Costs} - \text{Taxes} \tag{1.5}$$
$$\text{Cash Flows} = \$1,000,000 - \$775,000 - \$50,000 = \$175,000$$

or

$$\text{Cash Flows} = \text{Net Profits} + \text{Depreciation} \tag{1.9}$$
$$\text{Cash Flows} = \$150,000 + \$25,000 = \$175,000$$

Note that the difference in taxes which is called deferred taxes is:

$$\text{Deferred Taxes} = \text{Straight Line Taxes} - \text{Accelerated Taxes} \tag{1.10}$$
$$\text{Deferred Taxes} = \$50,000 - \$43,750 = \$6,250$$

The deferred taxes can be determined by the product of the depreciation differences times the tax rate:

$$\text{Deferred Taxes} = (\text{Accelerated Depreciation} - \text{Straight Line Depreciation}) \times (\text{Tax Rate}) \tag{1.11}$$

$$\text{Deferred Taxes} = (\$50,000 - \$25,000) \times (0.25) = \$6,250$$

$$\begin{aligned}\text{Cash Flow (CF) Difference} =&\text{Accelerated Depreciation CF} \\ &- \text{Straight Line Depreciation CF}\end{aligned} \tag{1.12}$$
$$\text{Cash Flow (CF) Difference} =\$181,250 - \$175,000 = \$6,250$$

The earnings per share difference can be determined can be determined by:

$$\begin{aligned}\text{Earnings/Share(Difference)} =& (1 - \text{Tax Rate}) \\ &\times ((\text{Accelerated Depreciation-Straight Line Depreciation})/(\text{Number of Shares})\end{aligned} \tag{1.13}$$

$$\text{Earnings/Share(Difference)} = (1 - 0.25) \times (\$50,000 - \$25,000)/10,000$$
$$= 1.875 \text{ \$/share}$$

In summary, the accelerated depreciation results in lower taxes paid and higher cash flows than the straight line depreciation method, but it also results in lower profits and lower share earnings. Thus, they use the accelerated depreciation method for reporting to the government and the straight line depreciation method for reporting to shareholders. Although this practice is legal, it does raise concerns about it being ethical.

In some instances, a company may have one project with negative cash flows but a net positive cash flow overall. This can occur when a company has several projects and a start-up project may be negative in its initial stages and other projects in the company may have positive cash flows resulting in a net positive cash flow. Construction projects often take long time periods and they are not paid until certain goals have been reached and thus may have negative cash flows until they are paid. A longer working capital cycle often results in negative cash flows for short periods. For example, when a housing developer builds a home, the developer does not recover the expenses until the home is sold and has a negative cash flow until payment is received for the home.

Hence, a reasonable amount of positive cash flows from operations is significant for three reasons [1].

1. Healthy cash flows can help a company meet its funding requirements internally rather than borrowing in a high cost environment; but for major capital expenditures borrowing is often necessary and advantageous for the company.

2. Having cash available permits the making of purchases more quickly and often at lower costs.

3. A company's ability to manage its debts indicates the efficiency and strength of the business to its customers, stockholders, and employees.

1.4   CASH FLOW DIAGRAMS

Cash flow diagrams are diagrams of the revenues and expenses over time. The abscissa (x-axis) represents time and the time between periods, which is usually constant such as one year, whereas the ordinate (y-axis) represents the amount of cash flow and is usually different for each time period. Cash flows are assumed to occur at the end of the period. There are three primary types of cash flow diagrams: (1) the basic cash flow diagram where the net cash revenues and net cash expenses are both shown; (2) the net cash flow diagram where the net incomes per period and net expenses per period are combined into a net overall cash flow per period; and (3) the cumulative cash flow diagram where the cumulative of the net cash flows are plotted. The most common diagram is the net cash flow diagram and the second most utilized is the cumulative cash flow diagram to illustrate the breakeven time. Since the cash flows are considered to occur at the end of the period, the time values on the diagram represent the end of the period. One should always make a cash flow diagram when possible in solving problems and in most cases the flows will be different for the various time periods.

We will consider an example with an initial investment of $10 which results in a net revenue stream of $6 for each time period (which could be the sum of two or more individual revenue streams) and a net expense stream of $3 over 5 time periods. Figure 1.1 shows the basic cash flow diagram, with the investment (an expense) occurring at time zero and the net revenues (positive) and net expenses (negative) for each period of the study.

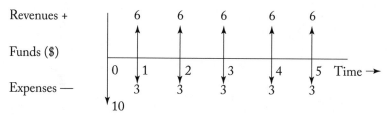

Figure 1.1: Basic cash flow diagram showing net revenue and net expense cash flows for each time period.

Figure 1.2 shows the net cash flows, the difference between the net revenues and net expenses. It is used as the starting point for calculations; somewhat like the use of the free-body diagram for solving in statics courses.

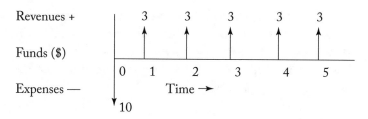

Figure 1.2: Net cash flow diagram showing net cash flows for each time period.

Figure 1.3 shows the cumulative cash flows which gives an indication as to when the product will become profitable, which is 4 years in this problem. If the cash flows were uniform over the project period, the result could be and is often interpreted as being 3.33 years. However, the cash flows are considered to occur at the end of the period and 4 years should be used if one does not know that the cash flows are uniform over the project period.

Since all cash flows are assumed to be at the end of the period, the initial investment occurs at time zero, which is the end of Period 0 and it is also the start of Period 1. Similarly, the end of Period 1 is also the start of Period 2. As we get into more complex examples, the cash flow per period will tend to be different for each period and the cash flow diagram helps in properly formulating the problem. Some of the other items that will be considered are accounts payable, accounts receivable, inventory changes, and etc., so some of the formulas presented must be adjusted for these items.

1.5 SUMMARY

Cash flows and profits are the two primary measures of corporate projects, but cash flow is the measure that is utilized for evaluating projects as positive cash flows are critical and profits are a

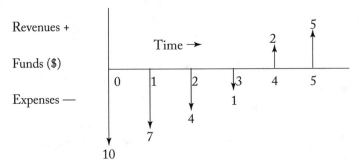

Figure 1.3: Cumulative cash flow diagram showing cumulative cash flow over each time period which indicates project becomes positive between the end of time period 3 and the end of time period 4.

component of the cash flows. Profits using accelerated depreciation are lower than the straight line depreciation, so accelerated depreciation is used for determining taxes and straight line depreciation is used for reporting to stockholders. The three cash flow diagrams were presented and the net cash flow diagram is used for most economic evaluations and the cumulative cash flow diagram is used for payback period calculations.

1.6   REFERENCES

[1] Creese, Robert C. and Adithan, M., *Strategic Cost Analysis for Project Managers and Engineers*, 2nd ed., New Academic Science Ltd., page 3, 2012. 8

1.7   EVALUATIVE QUESTIONS

1. Which depreciation method, accelerated or straight line, gives lower amounts of depreciation during the last stages of the project life?

2. Depreciation is considered to be what type of item and what is the main purpose of depreciation?

3. Use the information in Table 1.2 to determine the items listed below for the Accelerated Depreciation Method and for the Straight Line Depreciation Method. The company has 20,000 shares of stock.

   3.1 Use the Accelerated Depreciation method and determine:
      (a) Gross Profits
      (b) Net Profits

Table 1.2: Cash flow and depreciation data for problem 3

Item	Amount ($)	Amount (%)
Revenue	1,000,000	
Costs	800,000	
Tax Rate (%)		40
Depreciable Investment	400,000	
Depreciation—Straight Line	40,000	10
Depreciation—Accelerated (MACRS)	80,000	20

    (c) Taxes

    (d) Cash Flows

    (e) Net Profits/share

3.2 Use the Straight Line Depreciation method and determine:

    (a) Gross Profits

    (b) Net Profits

    (c) Taxes

    (d) Cash Flows

    (e) Net Profits/share

3.3 Determine the amount of deferred taxes and the difference in the net profits/share and the difference in cash flows between the Accelerated Depreciation and Straight Line Depreciation methods.

3.4 An error was made in determining the costs and instead of $800,000 they were $940,000. What does this difference make in the results? Calculate the Cash Flows and Net Profits/Share using both depreciation methods and compare the results.

4. Use the information in Table 1.3 to determine the items listed below for the Accelerated Depreciation Method and for the Straight Line Depreciation Method. The company has 20,000 shares of stock.

    4.1 Use the Accelerated Depreciation method and determine:

        (a) Gross Profits

        (b) Net Profits

        (c) Taxes

        (d) Cash Flows

        (e) Net Profits/share

Table 1.3: Cash flow and depreciation data for problem 4.

Item	Amount ($)	Amount (%)
Revenue	1,000,000	
Costs	800,000	
Tax Rate (%)		20
Depreciable Investment	500,000	
Depreciation—Straight Line	50,000	10
Depreciation—Accelerated (MACRS)	100,000	20

4.2 Use the Straight Line Depreciation method and determine:

(a) Gross Profits

(b) Net Profits

(c) Taxes

(d) Cash Flows

(e) Net Profits/share

4.3 Determine the amount of deferred taxes and the difference in the net profits/share and the difference in cash flows between the Accelerated and Straight Line Depreciation methods.

5. The large companies use accelerated depreciation when reporting to the government and straight line depreciation when reporting to stockholders.

5.1 This is legal, but is it ethical? Give your reasons for supporting your answer.

5.2 Should a law be made to permit the same depreciation method to both the stockholders and the government? Give your reasons for supporting your answer.

6. A company made an investment of $1,200,000 for a machine to manufacture a new product. The sale of the product produced is expected to provide a uniform annual revenue of $500,000 for 6 years. The annual operating, material, and maintenance expenses are $250,000 and the salvage value of the machine at the end of the 6 years is $400,000. Draw the cash flow diagram, the net cash flow diagram, and the cumulative cash flow diagram. What is the breakeven payback period?

7. Explain the difference between cash flows and profits with respect to depreciation.

8. (a) Given the following data in Table 1.4, create the basic cash flow diagram, the net cash flow diagram, and the cumulative cash flow diagram.

(b) When is the breakeven or payback period?

    (1) Assume end-of-year payments.

    (2) Assume uniform payments throughout the year.

(c) What is the total profit for the project?

Table 1.4: Data for problem 8

Period	Investment	Revenue	Expenses
0	18		
1		5	4
2		7	6
3		12	6
4		14	6
5		12	7
6		10	7
Totals	18	60	36

CHAPTER 2

Financial Statements and the Purcell Diagram

2.1 INTRODUCTION

Financial statements are critical in measuring the performance of an enterprise, and the two primary financial statements used are the Income Statement and the Balance Sheet. The Purcell Diagram [1, 2], developed by W.R. Purcell, integrates these two financial statements into a cash flow diagram. It indicates the overall cash flows for a company and illustrates the major components. This is critical for financial analysts and it assists engineers in understanding the importance of cash flows in the company picture and explains why project evaluations require a cash flow analysis.

2.2 FINANCIAL STATEMENTS

The primary financial statements used for reporting are the **Income Statement** and the **Balance Sheet**. The items such as long-term debt and short-term debt are not included to keep the problem easy to visualize and to reduce the complexity as they are difficult to add to the basic Purcell Diagram.

The income statement summarizes the revenues (sales), the major expenses (costs), the depreciation, the taxes, and the profits. Some items have been added to the expenses on the basic income statement to provide more detail on the total expenses. In the high technology society, more money is being spent on Research and Development (R&D) than in the past and these costs are significant. For example, as automobiles are switching from gasoline (fossil fuels) engines to electric motor engines, the development of these electrical engine systems are being done currently and the advanced models must undergo testing for performance and safety issues before they can be marketed. This R&D must be expensed during the development stage and the profits will occur only when the new technology is successfully marketed and those profits will be used to support future R&D developments. More advanced production equipment is being purchased to reduce labor and material costs and thus depreciation expenses will grow. In addition to the federal government taxes, other taxes and fees have been increasing and are added as a separate item. As labor costs are higher in the U.S., the top management costs have also increased significantly in salaries and stock options and the legal and computer security costs have also increased greatly due to cybersecurity threats. The example income statement for Shawnee Corporation with some of these components is presented in Table 2.1.

Table 2.1: Income statement for Shawnee Corporation in 2020

Income Statement		
Shawnee Corporation 2020 (End-of-Year)		
Sales		620
Expenses		
Cost of Goods Sold	290	
Management Costs	40	
R&D Expenses	60	
Sales Expense	70	
Other Taxes & Fees	20	
Depreciation	40	520
Profit Before Tax		100
Taxes (Tax Rate 25%)		25
Net Profit		75

The balance sheet indicates the financial positions of the company at the beginning and at the end of the accounting period to show the progress during the year. The basic equation of the balance sheet is that:

$$\text{Assets} = \text{Liabilities} + \text{Equities} \tag{2.1}$$

The relationship between cash flows and profits developed in Chapter 1 was:

$$\text{Cash Flows} = \text{Net Profits} + \text{Depreciation} \tag{2.2}$$

The more inclusive relationship between cash flows and profits is written as indicated by Equation (2.3) as:

$$\text{Cash Flows} = \text{Net Profits} + \text{Depreciation} + \text{Adjustments} \tag{2.3}$$

The adjustments to the cash flows are the accounts receivable, accounts payable, new equipment purchases, dividends paid, stock sales, and inventory changes and other items not included in the profit calculations. These adjustments can have a major impact on the cash flows and are considered in this chapter, but the long-term debt, short-term debt, principal, and interest payments are not included at this level of development. However, these can be included in the analysis in more advanced models. The balance sheet for Shawnee Corporation is shown in Table 2.2.

Table 2.2: Balance sheet for Shawnee Corporation in 2020

Balance Sheet Shawnee Corporation (Year 2020)					
	Start 2020	End 2020		Start 2020	End 2020
Assets			**Liabilities & Equities**		
Current Assets			**Current Liabilities**		
Cash	260	290	Accounts Payable	0	20
Accounts Receivables	0	40			
Inventories					
Finished Goods	50	25			
Work-in-Progress	45	80			
Raw Materials	30	30			
Fixed Assets			**Owner's Equity**		
Plant & Equipment	100	110	Common Stock	485	555
	485	575		485	575

2.3 THE PURCELL DIAGRAM

The Purcell Diagram gives the cash flows in a combined format of the Income Statement and the Balance Sheet [3]. The Purcell Diagram combines the information of the Income Statement and the Balance sheet into a single representation of the cash flows. The primary advantage of the Purcell Diagram is that it shows how the cash is flowing through the company and is more dynamic than the balance sheet and income statement. The Purcell diagram also uses more detailed data, such as the Purchased Equipment ($50), Stock Sales ($5), and Dividends Paid ($10), and illustrates more details of the owner's equity.

The Purcell Diagram shown in Figure 2.1 gives the cash flows in a much easier format than the balance sheet and is recommended for use. If one rewrites Equation (2.3) to determine the end-of-period cash flows, one has:

$$\text{Cash Flows} = \text{Net Profits} + \text{Depreciation} + \text{Adjustments} \tag{2.3}$$

$$\text{Cash Flows (end)} - \text{Cash Flows (start)} = \text{Net Profits} + \text{Depreciation} + \text{Adjustments}$$

or

$$\text{Cash Flows (end)} = \text{Cash Flows (start)} + \text{Net Profits} + \text{Depreciation} + \text{Adjustments} \tag{2.4}$$

From Tables 2.1, 2.2, and Figure 2.1, one can calculate the ending cash flows as:

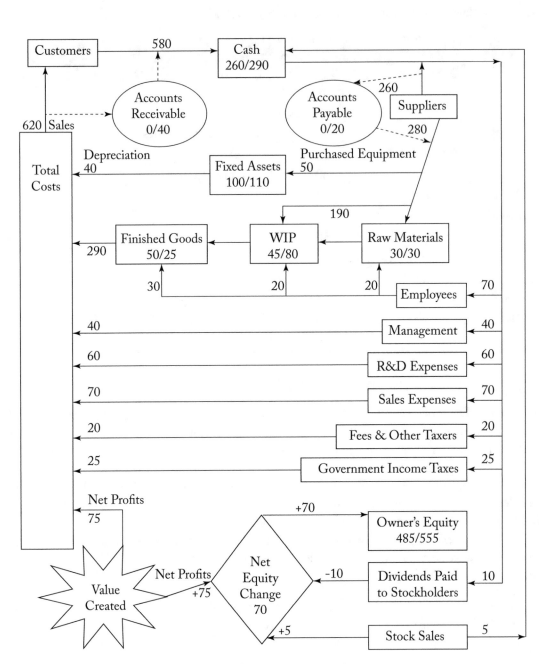

Figure 2.1: Purcell diagram for Shawnee Corporation in 2020.

$$
\begin{aligned}
\text{Cash Flows (end)} \quad &= \text{Cash Flows (start)} + \text{Net Profits} + \text{Depreciation} + \text{Adjustments} \\
&= +260 \qquad\qquad\qquad\quad + 75 \qquad\qquad\quad + 40 \\
&\qquad \{ - \text{Purchased Equipment} + \text{Stock Sales} \\
&\qquad \{ -50 \qquad\qquad\qquad\qquad\quad + 5 \\
&\quad - \text{Dividends Paid} - \text{Total Inventory Increase } (-25 +35 +0) \\
&\quad - 10 \qquad\qquad\qquad\quad\; -10 \\
&\quad + \text{Accounts Payable} - \text{Accounts Receivable} \} \\
&\quad +20 \qquad\qquad\qquad\quad -40 \,\} \\
&= 260 + 75 + 40 - 50 + 5 - 10 - 10 + 20 - 40 \\
&= 260 + 115 \;(\text{cash flow change via profits and depreciation}) \\
&\qquad\quad - 85 \;(\text{total adjustments}) \\
\text{Cash Flows (end)} \quad &= 290
\end{aligned}
$$

Note that the capital equipment is not listed as an expense item on the Income Statement and appears only on the Purcell diagram. Capital equipment must be depreciated over the life of the equipment and the depreciation is listed as an expense on the Income Statement. However, it does affect the cash flows and is presented as one of the adjustments to the cash flows. It is included in the fixed assets as the increase in the fixed assets represents the difference between the purchased equipment (an increase in assets) and the depreciation (a decrease in the assets).

This can also be used in another manner for calculating the ending value of a particular activity. The ending values need to be calculated and the general equation used is:

$$ \text{Ending Value} = \text{Starting Value} + \text{Inputs to Activity} - \text{Outputs of Activity} \qquad (2.5) $$

If one examines the cash flow activity in the Purcell Diagram, it is noted that:

$$ 290(\text{end}) = 260(\text{start}) + 580(\text{receipts from customers}) - 555(\text{outgoing funds}) + 5(\text{stock sales}) $$

Equation (2.5) can be used to determine all the ending values for the balance sheet and Purcell Diagram. The Purcell diagram shows the inputs and outputs for the various activities. The Purcell diagram is an excellent tool in complementing the Income Statement and the Balance Sheet.

2.4   SUMMARY

The Purcell diagram ties together the Income Statement and Balance Sheet to illustrate how the cash flows through the financial institution. It shows why cash flows are important and how they are related to profits, owner's equity, fixed assets, and the other key financial items for monitoring the performance of the financial institution. The focus of most of the remaining chapters will be on the cash flow calculations and the Purcell Diagram indicates how they relate to the financial statements of the Balance Sheet and Income Statement.

2.5   REFERENCES

[1] Purcell, W. R., *Understanding a Company's Finances—A Graphical Approach*, Houghton Mifflin Company, Boston, 1981. 15

[2] Purcell, W. R., *Understanding a Company's Finances: Look at Financial Reports*, see a Financial Picture of the Business, July 25, 2009, Kindle eBook. 15

[3] Creese, Robert C. and Adithan, M., *Strategic Cost Analysis for Project Managers and Engineers*, 2nd ed., New Academic Science Ltd., 2012. 17

2.6   EVALUATIVE QUESTIONS

1. Use the Income Statement and the Balance Sheet in Tables 2.3 and 2.4 to complete the Purcell Diagram for the Financial Flows for the Shawnee Corporation in Figure 2.2. The equipment purchased during the year was 70 and the labor was the same as 2020. Fill in the blanks in Table 2.4 and in the Purcell Diagram of Figure 2.2.

Table 2.3: Income statement for Shawnee Corporation for 2021

Income Statement		
Shawnee Corporation 2021 (End-of-Year)		
Sales		655
Expenses		
Cost of Goods Sold	285	
Management Costs	40	
R&D Expenses	60	
Sales Expense	70	
Other Taxes & Fees	20	
Depreciation	60	535
Profit Before Tax		120
Taxes (Tax Rate 25%)		30
Net Profit		90

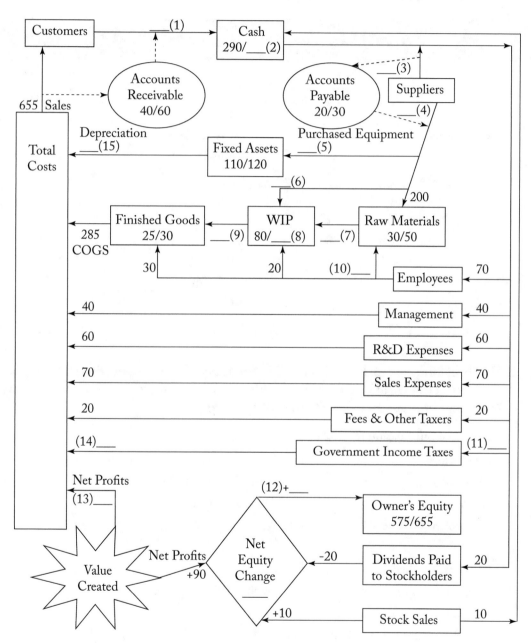

Figure 2.2: Purcell diagram for Shawnee Corporation in 2021.

Table 2.4: Balance sheet for Shawnee Corporation for 2021

Balance Sheet Shawnee Corporation (Year 2021)					
	Start 2021	End 2021		Start 2021	End 2021
Assets			**Liabilities & Equities**		
Current Assets			**Current Liabilities**		
Cash	290	__(1)	Accounts Payable	20	30
Accounts Receivables	40	60			
Inventories					
Finished Goods	25	30			
Work-in-Progress	80	85			
Raw Materials	30	40			
Fixed Assets			**Owner's Equity**		
Plant & Equipment	110	__(2)	Common Stock	__(4)	__(5)
	575	__(3)		575	__(6)

2. Use the information in Tables 2.5 and 2.6 with the additional information and complete the Purcell Diagram in Figure 2.3. The additional information given is for calculating some of the adjustment terms in Equations (2.3) and (2.4).

Additional information for 2024 is:

Stock Sales	20	Cash Received from Customers	780
Dividends Paid	20	Raw Materials to WIP	240
Labor Used		WIP to Finished Goods	325
For Raw Materials	40		
For WIP	35		
For Finished Goods	55		
Supplier Services		Money Paid to Suppliers	325
Equipment Purchased	90		
Raw Materials for Prod.	200		
WIP Services	45		

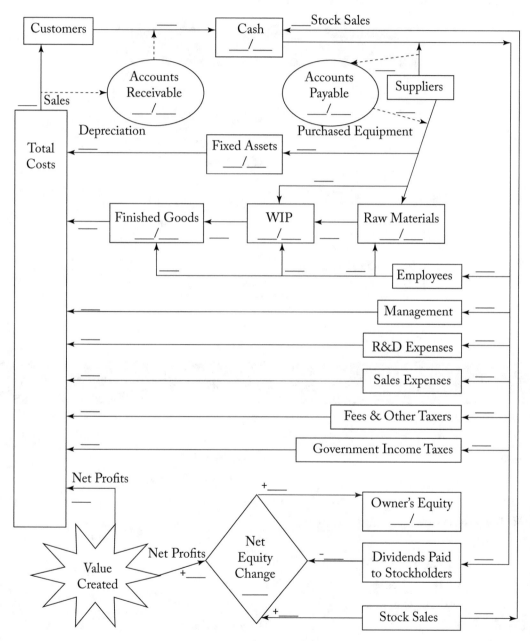

Figure 2.3: Purcell diagram for NAC Corporation in 2024.

Table 2.5: Income statement for NAC Corporation for 2024

Income Statement		
NAC Corporation 2024 (End-of-Year)		
Sales		800
Expenses		
Cost of Goods Sold	400	
Management Costs	40	
R&D Expenses	60	
Sales Expense	70	
Other Taxes & Fees	30	
Depreciation	80	680
Profit Before Tax		120
Taxes (Tax Rate 25%)		30
Net Profit		90

Table 2.6: Balance sheet for NAC Corporation for 2024

Balance Sheet					
NAC Corporation (Year 2024)					
	Start 2024	End 2024		Start 2024	End 2024
Assets			**Liabilities & Equities**		
Current Assets			**Current Liabilities**		
Cash	320	415	Accounts Payable	0	10
Accounts Receivables	0	20			
Inventories					
Finished Goods	40	20			
Work-in-Progress	90	85			
Raw Materials	40	40			
Fixed Assets			**Owner's Equity**		
Plant & Equipment	100	110	Common Stock	590	680
	590	690		590	690

CHAPTER 3

Costs and Cost Estimating

3.1 INTRODUCTION

Costs, interest payments, and cost estimating have been in existence since the first merchants started buying and selling products. The focus in business today tends to be on profits and cash flows, but the primary component in determining the amount of profits or cash flows are the expenses or costs. Some Biblical verses related to interest and cost estimating are presented to indicate that the issues of interest and cost estimating have existed for a long time; the vs. are paraphrased as:

1. **Deuteronomy 23: 19–20** You must charge no interest on a loan made to your brother. On a loan to a foreigner, you must charge interest…

2. **Luke 14:28** For which of you, wanting to build a tower, does not first sit down to determine the cost, to see if he has enough to complete it?

Although most religions now tend to charge interest to all people, a few countries/communities still tend to charge interest only to foreigners. Although costs, cost estimating, and interest payments are not new and have been in existence since pre-Biblical times, the methods for evaluating them have changed considerably over the centuries.

Two terms that are related and represent the before and after of a project/product are cost estimating and cost accounting. Cost estimating is done to determine if a project/product is feasible and cost accounting is to determine the actual costs of the project/product for evaluating the profits and determining the accuracy of the cost estimate. These two definitions are from AACE International Recommended Practice No. 10S-90, *Cost Engineering Terminology* [1], as revised October 30, 2017.

Cost Estimating: Cost estimating is the predictive process used to quantify, cost, and price the resources required by the scope of an investment option, activity, or project. Cost estimating is a process used to predict uncertain future costs. In that regard, a goal of cost estimating is to minimize the uncertainty of the estimate given the level and quality of scope definition. The outcome of cost estimating ideally includes both an expected cost and a probabilistic cost distribution. As a predictive process, historical reference cost data (where applicable) improve the reliability of cost estimating. Cost estimating, by providing the basis for budgets, also shares a goal with cost control of maximizing the probability of the actual cost outcome being the same as predicted (November 2012).

Cost Accounting: The historical reporting of actual and/or committed disbursements (costs and expenditures) on a project. Costs are denoted and segregated within cost codes that are defined in a chart of accounts. In project control practice, cost accounting provides the measure of cost commitment and/or expenditure that can be compared to the measure of physical completion (or earned value) of an account (January 2003).

The calculation of profits requires the determination of both revenues and expenses/costs. The expenses/costs represent the majority of the items on the Balance Sheet and management has more control over the expenses/costs than on the revenues items. The revenues have more outside factors affecting their control, such as customer demands, competition, and general economic prosperity. Cash flows are closely related to profits and both are essential in determining the success of a company.

To illustrate the influence of costs upon profits, Table 3.1 compares the effect of a 10% change in increased price, increased sales volume, and cost reductions upon the gross profits [2]. It indicates that the price increase is the best option, but it is only slightly better than the 10% cost reduction. It is much more likely for companies to achieve a cost reduction of 10% than an increase of selling price of 10%. With this given scenario, the 10% price increase results in a 125% increase in total profits and a 10% cost reduction results in 115% increase in profits. This is why companies focus on cost reductions when trying to increase profits as they have more control on costs whereas price increases are difficult to achieve in a competitive market. When there is a monopoly, outrageous price increases can occur, which has occurred in the medical industry when only one supplier exists and when one has her/his life involved. This is, however, is an exception to usual business practices, but it illustrates that unethical practices are often not illegal.

There are three basic approaches to cost estimating which are: Top-Down Estimating, Bottom-Up Estimating, and Combined Top-Down and Bottom-Up Estimating. There are other approaches to cost estimating, but these are the primary approaches utilized. Each of the approaches will be illustrated by examples, but the traditional method of cost estimating has been the bottom-up approach.

In the bottom-up estimating approach the complete details of the production process are known and cost for each step of the process can be estimated. The cost of all the steps of the process are estimated and the total cost of all the steps is the estimate for the product or project. This is done by most companies when the new products are similar to existing products and relatively small changes occur in the processes and the estimates tend to be accurate. Greater changes in the new product or process tend to result in more error in the estimate accuracy as items may be omitted in the estimate or more corrections will be needed in the manufacturing process. A foundry producing cast iron six-cylinder engine blocks could use that as the basis for making an estimate for a cast iron four-cylinder or an eight-cylinder engine block as most of the steps in the process would be similar. However, the changing from a six-cylinder cast iron

Table 3.1: Gross profit improvement analysis for a two product system [2]

	Current Practice	10% Price Increase	10% Sales Increases	10% Cost Reduction
Total Cost + Profit				
Product A	80	88	88	80
Product B	20	22	22	20
Total	100	110	110	100
Variable Costs				
Product A	57	57	62.7	51.2
Product B	12	12	13.2	10.8
Fixed Costs				
Product A	19	19	19	17.1
Product B	4	4	4	3.6
Total Cost				
Product A	76	76	81.7	68.4
Product B	16	16	17.2	14.4
Gross Profit				
Product A	4	12	6.3	11.6
Product B	4	6	4.8	5.6
Total	8	18	11.1	17.2
Profit Increase				
Amount				
Product A	Base	8	2.3	7.6
Product B	Case	2	0.8	1.6
Total		10	3.1	9.2
Profit Increase				
Percent				
Product A (%)	Base	200	57.5	190
Product B (%)	Case	50	20	40
Total Percent		125	38.8	115

engine to a six-cylinder aluminum would be more difficult as the process steps differences would be greater.

The traditional top-down estimating approach is usually based on Cost-Estimating Relationships (CERs) using primary cost drivers as the actual process steps and design details may not be known. These tend to focus on products with significant changes in design and where the manufacturing steps and sequence are not known. This often occurs for new military equipment to be manufactured and performance goals are specified, but the detailed product design has not been completed.

A second approach to top-down estimating is Target Costing, which is done in industries which purchase many of the components for their product. In the automobile industry, the manufacturers must estimate the total cost of the model to assure that it will be competitive in the market place. The automobile manufacturers produce the main components such as the engine, power train, frame, and skin, but usually have suppliers produce the tires, wheels, seats, wiring harnesses, glass, audio, etc. They need to set targets for these supplier components so that they can meet their total cost target to be competitive in the market place.

The third cost approach is a combination of top-down and bottom-up costing. This is done for projects of long duration, such as the production of an aircraft carrier. It takes several years to make the product, such as 10 years, and technology will change significantly over that time. The basic ship building costs can be estimated by bottom-up costing, but the cost for the electronics such as radar systems and defensive systems that will not be designed until several years in the future and thus these costs cannot be determined by bottom-up costing approaches, but will need CERs to estimate the costs.

3.2   COST COMPONENTS FOR ESTIMATES

3.2.1   BASIC COMPONENTS

There are numerous cost components and they will vary by industry, company and product and thus only a sample of the items commonly used in manufacturing are presented in Table 3.2. Other companies, such as financial services, food services, medical services, hotel and motel industries, and travel services will have similar components, perhaps with different names. In activities such as R&D, the expenses of the R&D can be determined by direct cost methods, but they must be allocated as an overhead to current products to recover the overhead costs.

3.2.2   TRADITIONAL AND ABC OVERHEAD ALLOCATION METHODS

The overhead costs are replacing direct labor costs as the largest cost component because direct labor costs have decreased through automation replacing labor and a large increase has occurred in overhead cost components such as R&D, legal, safety, and administrative management costs. The material cost proportion can vary considerably depending up the amount of purchased materials involved. In the automotive manufacturing business, most auto companies purchase the

Table 3.2: Cost components for estimating in manufacturing

A. Pre-production Costs (often in Corporate Overhead and not as separate costs)
    1. Research & Development
    2. Engineering Design
    3. Process Engineering Design (Tooling)

B. Production Costs (Direct)
    4. Direct Materials
    5. Direct Labor
    6. Other Direct Costs
       i. Tooling
       ii. Processing
    7. Contingency/Risk
       i. Product-Engineering Design Changes
       ii. Process-Tool Changes

C. Overhead Costs (Indirect)
    8. Overhead Costs
       i. Shop (Including Indirect Materials & Labor)
       ii. Plant
       iii.Corporate
       iv. Sales
    9. Adjustments to Costs
       i. Quantity Adjustments
       ii. Surcharges
          a. Special Testing
          b. Special Delivery

D. Total Costs
    10. Total Costs = Σ1 through 9 + Taxes

E. Profit
    11. Profit (Mark-up) + Tax Estimate on Profit

F. Selling Price
    12. Selling Price
       Selling Price = Total Costs + Profit + Estimated Profit Tax

glass, the interior seats, the tires, the head and tail light assemblies, the electrical harnesses, etc., so the total material costs are high as the purchased materials are the finished products of the suppliers. In companies that produce the glass for the auto windows, the raw materials are a relatively small portion of the total costs, but the processing costs are very high.

One approximate method for determining the selling price based on the material cost was the 1-3-9 rule of Rondeau [3]. The 1 represented the material cost (or 1.2 including scrap and tooling), the 3 representing the manufacturing cost, and the selling price represented by the 9. Thus, for a product that used materials of $3, the manufacturing cost would be $9 and the selling price would be $27. The manufacturing cost includes factory overhead including factory administration, product scheduling, quality control, material handling, shipping and receiving, as well as direct labor. The selling price includes the administrative overhead, the R&D, information systems and security (ISS), legal overheads, product testing, sales and marketing expenses, taxes, and a mark-up of 10–25%. The ratio can be developed for specific companies or industries, but the total overheads and other expenses are typically much greater than the manufacturing costs.

There are two major approaches to estimating of overhead costs. The traditional approach uses one or more of the direct costs to estimate the overhead cost, such as direct labor dollars, direct labor hours, or total direct costs. The Activity Based Costing (ABC) uses the measurement of a major activity for a department, such as purchasing orders, and determines a unit cost for that activity which can be assigned to a specific department or product. The determination of specific activities for administration, R&D, sales, and marketing are difficult to develop and this has been a problem with the traditional ABC.

The traditional allocation of overhead was based upon the amount of direct labor used in the production, which could be measured in either hours or dollars. This was fairly accurate when direct labor was largest cost component as in the early 1900s, but over the last 75 years companies have modernized equipment and used more automated material handling and robotics to reduce the amount of direct labor cost has been reduced from 40–60% of the total costs to 10–30%. New costs such as ISS, computer maintenance, and enhanced legal and safety expenses have increased over the years so labor is no longer the dominant cost as the total support costs far exceed the labor costs. Today direct labor costs are only 10–30% of the total costs and overhead rates based only upon direct labor would result in percentages from 300–500%. In many cases, these overheads are not even closely related to the amount of direct labor and this results in poor overhead allocations. This has resulted in other variables and/or additional variables being utilized to estimate the overhead costs. Direct costs, such as material cost, can be assigned easily to the products, but the support costs often have little direct connection to the various individual products.

ABC takes the cost of the overhead activity and divides by the performance measure for that activity to determine the rate. The overhead charge to the consuming department is the product of the rate determined and the amount of the activity consumed by the department. For example, the purchasing department is a service activity and its expenses are overhead. Thus, the expenses of the department divided by the number of purchase orders processed would be the rate and the production department would be charged by the product of the number of purchase orders submitted times the rate. This led to several problems in determining rates as one purchase order may contain only one item and another purchase order would contain numerous items. Also, many of the purchase orders were for other support groups and could not be related to specific products.

The newer approach is to develop rates based on the time for the activity and this has been used in both ABC and traditional overhead determination. Time is what is being consumed and the specific time values closely related to production are the direct labor time and the time machines are being used for production. The use of time has changed the traditional ABC to Time-Based ABC with the use of activities being related to time. The overhead costs are a much larger component of the total cost and they are different in their support of the product, so different measures are needed and they can be measured in units, such as units of production or in time units such as machine hours, direct labor hours, or the sum of machine and labor hours. Eventually the overheads need to be assigned to specific products to determine the product cost and a selling price for recovering the costs and providing an acceptable profit.

An example is presented in Table 3.3 of a metal casting facility which is a supplier to the automotive industry. The traditional ABC will be used for two activities and the administrative and facility overheads will be determined by using the traditional direct labor hours, the machine hours and the sum of the direct labor and machine hours. There are four products—pistons, crankshafts, axles, and exhaust manifolds—and the unit selling price, unit direct material costs, direct labor hours, and machine hours, as well as the ABC data for purchase orders and material handling are presented. The machine hours are used for both the depreciation unit and utility unit charge determinations.

The quantities, selling prices, and direct costs are known, and the problem is to determine how to allocate the factory and administrative overhead expenses of $900,000 and the other overheads of $652,000 to determine the individual product profit/loss. The ABC method utilizes two drivers in the determination of the overhead and they were: Number of Production Orders and Number of Moves. The Number of Machine Hours would be considered as a driver in either the ABC or Traditional Method. The Number of Machine Hours and the Number of Labor Hours are used for direct costs as they can be directly associated with a product. The data of Tables 3.3 and 3.4 are used to determine the unit cost values for the four products.

Table 3.3: Cost data for overhead evaluations for an automotive metal casting facility

| Overhead Item/ Activity | Cost Driver | Casting Products Produced | | | | Total # of Activity | Total OH Cost - $ | OH Rate ($/unit) |
		Piston	Crank-shaft	Rear Axel	Exhaust Manifold			
Purchase Orders	# of Orders	120	300	200	380	1,000	56,000	56
Material Handling	# of Moves	200	350	450	400	1,400	56,000	40
Depreciation Expense	# of Machine Hours	2,000	4,000	2,000	2,000	10,000	500,000	50
Utility Costs	# of Machine Hours	2,000	4,000	2,000	2,000	10,000	40,000	4
Factory Expenses							500,000	
Administrtive Expenses							400,000	
Total Expenses							1,552,000	

Table 3.4: Unit cost determination

| Production Data | Casting Products Produced | | | | Total # of Activity | Total OH Cost - $ | OH Rate ($/unit) |
	Piston	Crank-shaft	Rear Axel	Exhaust Manifold			
Production Units	38,000	8,000	6,700	34,000			
Selling Price: $/unit	10	132	75	23			
Direct Labor: hours	1,000	5,000	2,000	5,000	13,000	520,000	40
Utility Cost: Machine hours	2,000	4,000	2,000	2,000	10,000	40,000	4

Direct Unit Costs and Overhead Activities Directly Related to Products					
Production Costs		Piston	Crank-shaft	Rear Axel	Exhaust Manifold
Direct Materials	$/unit	4.000	32.000	19.000	6.000
Direct Labor	$/unit	1.053	25.000	11.940	5.882
Purchase OH	$/unit	0.177	2.100	1.672	0.626
Material Handling	$/unit	0.211	1.750	2.687	0.471
Depreciation Expenses	$/unit	2.632	25.000	14.925	2.941
Machine Utility	$/unit	0.211	2.000	1.194	0.235
Total Direct Costs	$/unit	8.284	87.850	51.418	16.155

The calculations for the unit costs for the Pistons in Table 3.4 will be illustrated.

The Direct Material Cost per unit is given as:

$$DMC = \$4.00/\text{unit}$$

The Direct Labor Cost per unit is calculated by:

$$DLC = (\text{direct labor overhead rate} \times \text{piston direct labor hours})/\text{\# of pistons produced} \tag{3.1}$$

$$= 40 \text{ \$/hr} \times 1{,}000 \text{ hrs}/38{,}000 \text{ pistons} = \$1.053/\text{unit}$$

The Purchase Order Cost per unit is calculated by:

$$POC = (\text{\# of Piston Purchase Orders} \times \text{Purchase Order OH rate})/\text{\# of Pistons produced} \tag{3.2}$$

$$= (120 \text{ purchase orders} \times \$56/\text{purchase order})/38{,}000 \text{ pistons} = \$0.177/\text{unit}$$

The Material Handling Cost per unit is calculated by:

$$MHC = (\text{\# of Piston Material Moves} \times \text{Material Handling OH rate})/\text{\# of pistons produced} \tag{3.3}$$

$$= (200 \text{ Piston Material Moves} \times \$40/\text{purchase order}/38{,}000 \text{ pistons} = \$0.211/\text{unit}$$

The Depreciation Expense Cost per unit can be calculated by:

$$DEC = (\text{\# of Piston Machine Hours} \times \text{Depreciation Expense OH rate})/\text{\# of pistons produced} \tag{3.4}$$

$$= (2{,}000 \text{ Piston Machine Hours} \times \$50/\text{machine hour})/38{,}000 \text{ pistons} = \$2.632$$

The Machine Utility Cost per unit is also based upon machine hours and can be calculated by:

$$MUC = (\text{\# of Piston Machine Hours} \times \text{Utility Cost OH rate})/\text{\# of pistons produced} \tag{3.5}$$

$$= (2{,}000 \text{ Piston Machine Hours} \times \$4/\text{machine hour})/38{,}000 \text{ pistons}) = 0.211$$

The sum of these costs are presented in Table 3.4 and are the sum of the material cost and the five calculated overhead costs as:

$$\text{Total Direct Assignable Costs (TDAC)} = \text{DMC} + \text{DLC} + \text{POC} + \text{MHC} + \text{DEC} + \text{MUC} \tag{3.6}$$

$$\text{TDAC} = 4.000 + 1.053 + 0.177 + 0.211 + 2.632 + 0.211$$
$$= 8.284$$

The sum of these direct and overhead costs is \$8.284, but this does not include the factory overhead expenses and administrative expenses. The unit costs for the other three products—Crankshaft, Rear Axle, and Exhaust Manifold—can be evaluated in a similar fashion to that presented for the Pistons. The factory and administrative expenses are related to time as many are salaries and benefits and the labor time and/or machine time are logical predictors for allocating these overhead expenses. The considerations presented are labor time, machine time, and the combined labor and machine time. Historical data should be used to select the best time base to use to allocate the Factory Overhead, Administrative Overhead, and other overheads such as R&D and Sales and Marketing. These overhead parameters may be different for the various overheads, that is Factory Overhead may be related best with direct labor hours while Administrative overhead may best be related to total labor plus machine hours. Table 3.5 presents the allocation of the Factory Overhead and Administrative Overhead first as a function of direct labor hours, then with machine hours, and finally with the sum of labor and machine hours.

The factory and administrative overheads will be calculated for direct labor hour basis to illustrate the calculations involved for the Piston.

The administrative overhead (PAOH) for the Piston in \$/Piston is calculated as:

PAOH = Piston Administrative Overhead

$$\text{PAOH} = (\text{Piston DL Hours/Total DL Hours}) \times (\text{Total Administrative Expenses/Total Pistons}) \tag{3.7}$$

$$= (1{,}000 \text{ hrs}/13{,}000 \text{ hrs}) \times (\$400{,}000/38{,}000) = 0.8097$$

The factory overhead (PFOH) for the Piston in \$/Piston is calculated as:

PFOH = Piston Factory Overhead

$$\text{PFOH} = (\text{Piston DL Hours/Total DL Hours}) \times (\text{Total Factory Expenses/Total Pistons}) \tag{3.8}$$

$$= (1{,}000 \text{ hrs}/13{,}000 \text{ hrs}) \times (\$500{,}000/38{,}000) = 1.0121$$

Table 3.5: Total product cost using different time basis

Direct Labor Hour Basis for Factory and Administrative Overhead Unit Costs					
		Piston	Crank-shaft	Rear Axel	Exhaust Manifold
Total Direct Costs	$/unit	8.284	87.850	51.418	16.155
Factory OH	$/unit	1.012	24.038	11.481	5.656
Adminisstrative OH	$/unit	0.810	19.231	9.185	4.525
Total Product Cost	$/unit	10.106	131.119	72.084	26.336
Selling Price	$/unit	10.000	132.000	75.000	23.000
Product Profit/Loss	$/unit	-0.106	0.881	2.916	-3.336
Machine Hour OH Basis for Factory and Administrative Overhead Unit Costs					
		Piston	Crank-shaft	Rear Axel	Exhaust Manifold
Total Direct Costs	$/unit	8.284	87.850	51.418	16.155
Factory OH	$/unit	2.632	25.000	14.925	2.941
Adminisstrative OH	$/unit	2.105	20.000	11.940	2.353
Total Product Cost	$/unit	13.021	132.850	78.284	21.449
Selling Price	$/unit	10.000	132.000	75.000	23.000
Product Profit/Loss	$/unit	-3.021	-0.850	-3.284	1.551
Total of Machine plus Labor Hour OH Basis for Factory and Administrative Overhead Unit Costs					
		Piston	Crank-shaft	Rear Axel	Exhaust Manifold
Total Direct Costs	$/unit	8.284	87.850	51.418	16.155
Factory OH	$/unit	1.716	24.457	12.979	4.476
Adminisstrative OH	$/unit	1.373	19.565	10.383	3.581
Total Product Cost	$/unit	11.373	131.872	74.779	24.212
Selling Price	$/unit	10.000	132.000	75.000	23.000
Product Profit/Loss	$/unit	-1.373	0.128	0.221	-1.212

The total unit product cost is:

Total Unit Piston Cost (TUPC)

$$\begin{aligned} &= \text{Total Direct Assignable Piston Costs (TDAPC)} + \text{Piston Administrative} \\ &\quad \text{Overhead (PAOH)} + \text{Piston Factory Overhead (PFOH)} \\ &= 8.284 + 0.810 + 1.012 \\ &= 10.106 \ \$/\text{piston} \end{aligned} \quad (3.9)$$

The total profit would be the unit selling price minus the unit total cost which for the piston is:

$$\text{Total Profit for Piston} = \text{Piston Selling Price/unit} - \text{Piston Total Cost/unit} \qquad (3.10)$$
$$= 10.000 - 10.106 = -0.106 \; \$/\text{piston (loss)}$$

This example problem is to show the methodology of the calculations. Note that the two other methods of assigning overheads gave pistons larger losses, the largest loss was $-3.019 using machine hours as the base for the factory and administration overheads. Pistons would require some precision machining and that would be a cause. This also suggests that the selling price is too low. In the automotive industry, the competition is fierce with the auto companies driving supplier prices down, but they do realize that they need the suppliers because they cannot make all the components at competitive costs.

Table 3.5 indicates that the rear axle has the highest profit amount using direct labor hours of $2.916/axle, but when using the machine hour base it has the greatest loss of $-3.284/axle. This variation indicates why it is important to determine which allocation base is best for allocating the factory expense and administrative overheads. One could also use different overhead bases, such as using direct labor hours for the factory expense allocation and machine hours for the administrative expense allocation. A previous work [4] has additional examples on overhead allocation.

3.2.3   PROFIT CALCULATIONS

If you don't know the costs, it is difficult to determine a competitive price, and nearly impossible to predict the profitability of the product. Price is based upon cost, but also includes other factors such as market conditions and the value of the product to the customer. Profit should **not** be a constant percentage or constant amount for all products, but should vary with respect to your ability to produce products. If you are "the best" on certain products, the profit should be greater on those products and lower on those products for which you are not "the best."

There are two methods commonly used to calculate the amount of profit to include in the total cost calculation. The two methods are the percent of cost and the percent of selling price and the difference is significant. An example of a product costing $100 will add 25% mark-up for profit.

Percent of Cost Calculations

$$\text{Selling Price} = \text{Product Cost} + \text{Product cost} \times \text{decimal percent mark-up} \qquad (3.11)$$
$$\text{Selling Price} = \$100 + \$100 \times 0.25$$
$$= \$100 + \$25$$
$$= \$125$$

The profit would be $= \$125 - \$100 = \$25$

Percent of Selling Price

$$\text{Selling Price} = \text{Product cost}/(1.0 - \text{decimal percent mark-up}) \qquad (3.12)$$
$$\text{Selling Price} = \$100/(1.00 - 0.25)$$
$$= \$100/(0.75)$$
$$= \$133.33$$

The profit would be $= \$133.33 - \$100 = 33.33$

Now since one is making 25% profit, a very good customer wants a 20% discount and the seller would still expect to make a 5% profit. The discount is applied to the selling price, not to the cost. Therefore, using the percent of cost calculations, the discount amount would amount to 20% of $125 which is $25 and thus the seller would have zero profit instead of 5% profit. If one uses the percent of selling price, the discount would be 20% of 133.33 which would be $26.66 and the seller would still have made a profit of $6.66 after the 20% discount, which is 5% of the $133.33. This is why the preferred mark-up approach is to base it upon selling price rather than cost.

Profit can also be considered on an after tax basis by including the estimated taxes as part of the selling price. Consider the previous example with a product cost of $100, a desired mark-up of 25% and an expected tax rate of 20%. What should the selling price be to obtain the desired mark-up of 25% after taxes using the Percent of Selling Price Method?

The profit after taxes as calculated above would be $= \$33.33$

The total profit needed including the taxes (Gross Profit) $= \$33.33/(1 - 0.20) = 41.666$

The selling price would be $= \$100 + \$41.66 = \$141.666$

The taxes would be 20% of the gross profit $= 0.20 \times 41.666 = 8.33$

The net profit is $\$141.66 - 8.33 - 100 = \$33.33$

3.3   COST ESTIMATION ACCURACY

The accuracy of estimates will vary considerably depending upon the amount and the accuracy of the information and the degree of complexity, knowledge, and experience of the estimators. AACE International has developed an estimating system [5] with five classes. The system presented uses that data, but has modified it giving two accuracy ranges based on two degrees of Difficulty/Complexity expressed as Low and High. The easy estimates (low difficulty) will have a higher percentage of the total estimating information required than the difficult, complex project estimates (high difficulty) and the accuracy ranges of the easy estimates would be smaller. However, the total amount of information required for the easy estimates is much less than the total amount of information required for the complex difficult projects and thus more information is required for complex projects than for the easy projects for all estimate classes. The primary factor is the known amount of information required to produce the product/project vs. the total amount of information required. Table 3.6 presents the expected accuracy range for an estimate based upon the estimate class, the percent of total estimating information required, and the difficulty level of the estimate. For example, consider a Class 4 estimate where a simple project may require 10 items of information and a complex project may require 200 items of information. If the simple project has one item of information and the complex project has 16 items of information, the percentages of total information required are 10% for the simple project and only 8% for the complex project. Thus, simple projects have a higher percentage of information required, but the complex projects require more information.

   If a Class 2 estimate is made for a complex, difficult project and has an estimated value of $3,000,000, what is the expect range of the total project cost?

   For a Class 2 estimate of a project with a high degree of difficulty, the estimate range is −15% to +20%. That implies the estimate could vary from 85–120% of the project estimate which is a range from $2,550,000–$3,600,000 for a total variation range of $1,050.000. Although the percentages of the estimate range seem small, for expensive projects the range is large in terms of dollars and that is why accurate estimates are important. For a Class 2 project with low difficulty, the range of the total project is −5% to +5%, which would result in a range of $2,850,000–$3,150,000 and a total variation range of $300,000 which is less than 30% of the high difficulty project range.

   Cost control is a strategic element in the success of an enterprise. Costs have a major impact upon profits and cost is the major item over which management has control in the amount of profits earned. With the reduction in direct labor costs, overhead costs have increased significantly and are a major element in the total costs and are more difficult to assign to products or projects than direct costs. Accurate estimates are necessary to provide information to determine the potential financial success of a product or project. The accuracy for estimates of small, similar products with minor differences have narrower ranges than complex projects such as the Mars Mission or newly designed complex products or projects. Cost engineers should monitor their estimates to develop an accuracy distribution and range for their estimates which may be different than those presented.

Table 3.6: Cost estimate accuracy based upon Information Required [5]

Estimate Class	Percent of Total Estimating Information Required (EIR (%)) and Difficulty Level (DL)		Typical Application	Expected Accuracy Range Based upon Project Difficulty/ Complexity	
	EIR (%)	DL		Accuracy Range (%)	
Class 5	1–10	Low	Screening	Low	-20 – +30
	0–2	High	Concept	High	-50 – +100
Class 4	5–20	Low	Budget Prep.	Low	-15 – +20
	1–15	High	Feasibility	High	-30 – +50
Class 3	15–50	Low	Budget Approval	Low	-10 – +10
	10–40	High	Budget Preparation	High	-20 – +30
Class 2	40–80	Low	Control Budget	Low	-5 – +5
	30–70	High	Budget Approval	High	-15 – +20
Class 1	75–100	Low	Check Estimate	Low	-3 – +3
	70–100	High	Bid/Tender	High	-10 – +15

*Table 3.6 is altered slightly from the original publication by AACE International [5]. This will reflect rearranging data of the last column to be two rows according to project difficulty/complexity level which originally was:

Low	-20% – -50%	-15% – -30%	-10% – -20%	-5% – -15%	-3% – -10%
High	+30% – +100%	+20% – +70%	+10% – +30%	+5% – +20%	+3% – +15%

3.4   SUMMARY

Costs and cost estimating are extremely important in the determination of cash flows and profits. The two major components are direct costs and overhead costs. The overhead costs have increased greatly due to safety, environmental, legal, and other new costs while the direct labor costs have decreased significantly over the past decades as a percentage of the total cost. The approaches to overhead allocation are the traditional methods which utilized percentage relationships of direct costs and the ABC method. The newer approaches are relating the costs to a time basis in both the traditional approach and Time-Based ABC method. If production time can be reduced, the overhead charge per unit will be reduced. The method of determining the product mark-up should be based upon a percentage of the selling price rather than a percentage of the total cost. The estimating class system permits a calculation of the range in the estimate based upon the complexity of the project and the experience of the estimators.

3.5   REFERENCES

[1] *Cost Engineering Terminology* AACE International Recommended Practice No. 10S-90, pages 29, 32, 2017. Copyright© 2017 by AACE International: All rights reserved. (Recommended Practice No. 10S-90, 2017, *Cost Engineering Terminology* is a free download to the public by visiting `web.aacei.org`) 25

Reprinted with the permission of AACE International, 1265 Suncrest Towne Centre Dr., Morgantown, WV, 26505, U.S. Phone 304-296-8444.
Internet: `http://web.aacei.org`
e-mail `mailto:info@aacei.org`

[2] Creese, Robert C., Adithan, M., and Pabla, B. S., *Estimating and Costing for the Metal Manufacturing Industries*, Marcel Dekker, Inc., page 13, 1992. (Reprinted with Permission of Taylor and Francis Group LLC Books.) 26, 27

[3] Rondeau, H. F., The 1-3-9 rule for product cost estimation, *Machine Design,* pages 50–53, August 1975. 30

[4] Creese, Robert C., Adithan, M., and Pabla, B. S., *Estimating and Costing for the Metal Manufacturing Industries,* Marcel Dekker, Inc., pages 35–49, 1992. 36

[5] *Cost Estimate Classification System—As Applied in Engineering, Procurement, and Construction for the Process Industries*, AACE International Recommended Practice No. 18R-97, pages 2–3, 2011. 38, 39

Reprinted with the permission of AACE International, 1265 Suncrest Towne Centre Dr., Morgantown, WV, 26505, U.S., Phone 304-296-8444.
Internet: `http://web.aacei.org`
e-mail: `mailto:info@aacei.org`

3.6   EVALUATIVE QUESTIONS

1. Find two additional references to interest charges or cost estimating before time zero AD.

2. In competitive markets, why is cost control performed more by management than increasing prices?

3. What are the major differences between cost estimating and cost accounting?

4. What are the three basic approaches to cost estimating?

5. Repeat the analysis of Table 3.1 for a 5% increase in price, a 5% increase in sales, and 5% cost reduction.

6. Determine the new overhead allocations for the Purchase Orders if the total overhead cost for the purchasing department is $60,000 and the orders for the products are: Piston-300, Crankshaft-500, Rear Axle-100, and Exhaust Manifold-100. Determine the unit purchasing OH cost for each of the four product produced and compare them with those in Table 3.4. The total quantities are the same as those listed in Table 3.4.

7. Determine the unit costs and total unit cost for the crankshaft if 10,000 units were produced (a recording error was made and 10,000 were made instead of 8,000) and compare the results (costs and profits) to the 8,000 total and unit costs of Tables 3.4 and 3.5. Assume all other values remain the same as in Tables 3.3, 3.4, and 3.5.

8. A product has a total cost of $2,000 and the desired profit is 15%.

   (a) Determine the selling price if the profit percent is based upon the cost.

   (b) Determine the selling price if the profit percent is based upon the selling price.

   (c) Determine the selling price if the profit percent is based upon the selling price and the tax rate is 20% and the profit desired is after taxes.

9. A Class 4 estimate is to be prepared for a project estimate of $1,500,000 that is very similar to previous work. What is the estimate range for this project?

10. Company JEN does complex projects and the estimated project cost is $5,000,000 for a Class 1 Estimate. What is the estimate range for this project and what is the minimum percentage of the total estimating information required?

CHAPTER 4

Breakeven Analysis

4.1   INTRODUCTION

The previous chapter indicated that the best approach for allocating overheads was the time-based system for either the traditional direct cost system or the Time-Based ABC system. Breakeven analysis has traditionally focused upon production quantity-based breakeven analysis and the cost breakeven point. This worked well for marketing, sales, and top-management for planning goals, but it provided little assistance at the plant management level where the production quantity is not a variable, but a quantity specified by the customer. The plant superintendent, production manager, or manufacturing manager can control the time to produce the orders but they cannot control the quantity. Thus, time-based breakeven analysis [1–5] is a concept being considered for use at the operation levels of production.

In addition to the two approaches to breakeven analysis, one can also consider different breakeven points and these will be presented in detail. The costs will be considered as fixed, variable, and semi-variable, but they will be considered differently for the two models as an item fixed in one approach may be variable in the other. For example, in the quantity-based system materials would be considered as variable whereas in the time-based system they would be considered as fixed as the quantity is fixed.

4.2   BREAKEVEN MODEL BASICS

The basis of the two models is the same equation presented by George Dieter in Volume 20 of the *ASM Handbook* [6]. The base equation is:

$$C_u = C_m + C_c/n + C_l/\text{pr}, \qquad (4.1)$$

where

C_u = unit cost, $/unit
C_m = unit material cost, $/unit
C_c = capital cost, $
C_l = labor and overhead cost, $/hr
n = production quantity, number of units
pr = production rate, units/hr

If one multiplies Equation (4.1) by the production quantity, n, to obtain the total cost, the equation for total cost (C_T) is:

$$C_T = nC_m + C_c + C_l n/\text{pr}. \tag{4.2}$$

Rewriting this equation with n being emphasized:

$$C_T = n \times (C_m + C_l/\text{pr}) + C_c. \tag{4.3}$$

If the production rate is considered to be a constant, the equation would be written as:

$$C_T = n \times (\text{slope constant in \$/unit}) + \text{intercept (\$)}$$

and this is the basis of the production quantity-based approach.

The basis of time-based approach is also from Equation (4.2)

$$C_T = nC_m + C_c + C_l n/\text{pr}. \tag{4.2}$$

If one recognizes that:

$n/\text{pr} = $ the total production time T and

$nC_m = $ the total materials cost for the order, C_{mt}

then with slight rearranging

$$C_T = T \times C_l + (C_{mt} + C_c). \tag{4.4}$$

Thus,

$$C_T = T \times (\text{slope constant in \$/unit time}) + \text{intercept (\$)}$$

and this is the basis of the time-based approach.

4.3   BREAKEVEN POINTS

The four breakeven points are defined so that they can be used in either the time-based system or the quantity-based system. The costs will be defined in three categories as fixed costs, variable costs and as semi-variable costs, but the specific costs may not be in the same category in the two bases. The typical costs in three categories will be listed for each cost base.

4.3.1   CATEGORIES AND TYPICAL EXAMPLES FOR THE PRODUCTION QUANTITY-BASED SYSTEM

In the production quantity-based system the overhead costs are not a separate category and are typically included in the fixed costs or may be assessed to the variable direct labor and/or direct material costs. The production costs will be considered in the three components of fixed costs, variable costs, and semi-variable costs.

- Fixed Costs—those costs which are independent of the production quantity required to make the product. The typical examples of fixed costs are property taxes, depreciation, administrative salaries, and plant overhead. The fixed overhead costs are often converted to be components of the direct labor cost.

- Variable Costs—those costs that are a direct function of production quantity to make the product. Two typical examples of variable costs are direct labor and direct material costs.

- Semi-Variable Costs—those costs which are partially fixed and partially variable. Costs such as maintenance expenses and inspection costs frequently are considered as semi-variable costs.

4.3.2   CATEGORIES AND TYPICAL EXAMPLES FOR THE PRODUCTION TIME-BASED SYSTEM

These costs are divided into two major groups as production costs and overhead costs. The production costs are divided into the three groups of fixed costs, variable costs, and semi-variable costs. Since this is a time-based system, the overhead costs can be considered as a separate category as a variable component.

a. Production Costs

   – Fixed Costs—those costs which are independent of the production time of the product; these would include the material costs. Since the quantity is fixed, the total material costs would be fixed.

   – Variable Costs—those costs which are directly dependent upon the production time of the product; these would include machine time costs, depreciation costs, plant overhead, and direct labor costs.

   – Semi-variable Costs—those costs which are partially fixed and partially variable; these include maintenance costs and utility costs.

b. Overhead Costs

   – Variable Costs—those costs which are dependent upon time but are not directly attributable to a specific product; these include administrative costs, research and development costs, etc.

Two example problems which have been presented previously [4] will be used to illustrate the types of variables, calculations, results, and interpretation of the results.

4.4   PRODUCTION QUANTITY-BASED BREAKEVEN EXAMPLE

The production quantity-based method is illustrated first as this is the traditional approach to breakeven analysis, but all four breakeven points will be illustrated. This is the same example used at the ASEE [4] conference and one should notice the differences in the data used for the two approaches. The data for the production quantity-based model is listed in Table 4.1.

Table 4.1: Data for production quantity-based breakeven analysis [1, 4]

Item	$/Unit	$	Decimal
Sales Revenue	20		
Production Costs			
Fixed Costs		2,400	
Variable Costs	3		
Semi-variable Costs	2	600	
Required Return (Profit)		900	
Tax Rate (40%)			0.40

The calculations for the four breakeven points are presented in a general form using the data from Table 4.1.

Let x = the units of production.

(a) Shutdown Breakeven Level (SD)

$$\text{Revenues} = \text{Variable Costs} + \text{Semi-variable Costs}$$
$$20x = 3x + 2x + 600$$
$$20x = 5x + 600$$
$$15x = 600$$
$$x = 40 \text{ units}$$

(b) Cost Breakeven Level (C)

$$\text{Revenues} = \text{Total Costs}$$
$$\text{Revenues} = \text{Variable Costs} + \text{Semi-variable Costs} + \text{Fixed Costs}$$
$$20x = 3x + 2x + 600 + 2,400$$
$$20x = 5x + 3,000$$
$$15x = 3,000$$
$$x = 200 \text{ units}$$

(c) Required Return Breakeven Level (RR)

$$\text{Revenues} = \text{Total Costs} + \text{Required Return}$$
$$\text{Revenues} = 5x + 3,000 + 900$$
$$20x = 5x + 3,900$$
$$15x = 3,900$$
$$x = 260 \text{ units}$$

(d) Required Return After Taxes Breakeven Level (RRAT)

$$\text{Revenues} = \text{Total Costs} + \text{Required Return After Taxes}$$
$$+ \text{Taxes on Total Required Return}$$
$$\text{Revenues} = \text{Total Costs} + \text{Required Return After Taxes}/(1.0 - \text{Tax Rate})$$
$$\text{Revenues} = 5x + 3,000 + 900/(1 - 0.4)$$
$$20x = 5x + 3,000 + 1,500$$
$$20x = 5x + 4,500$$
$$15x = 4,500$$
$$x = 300 \text{ units}$$

Note that to obtain a required return of $900 after taxes that one must earn $1,500 as the 40% taxes would be $600. To obtain the pre-tax required return, one can use the expression:

$$\text{Required Return Including Taxes} = \text{Required Return After Taxes}/(1.0 - \text{Tax Rate}) \quad (4.5)$$

The tax rate is expressed as a decimal.

Figures 4.1 and 4.2 show the breakeven points of total costs vs. production quantity and unit costs vs. total costs.

The results and actions for the various breakeven points are presented in Table 4.2. The Total Cost vs. Production Quantity is the breakeven figure observed, but the Unit Cost vs. Production Quantity shows the effects of revenue changes upon the various breakeven points more easily than that of Total Costs in Figure 4.1.

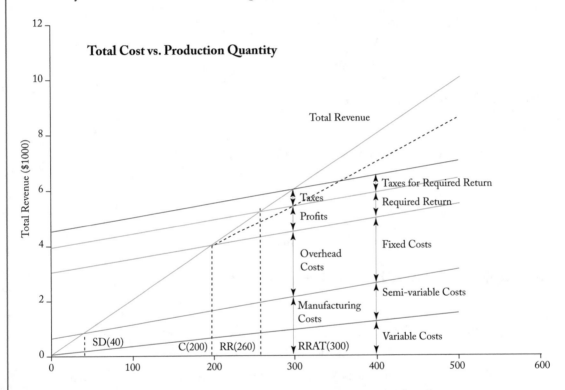

Figure 4.1: Total revenue and costs for the production-based model [4, 5].

The problem is that the customer usually dictates the level of production and the production department has little control over the quantity. The unit cost curve indicates what the revenue is required to obtain the desired breakeven points.

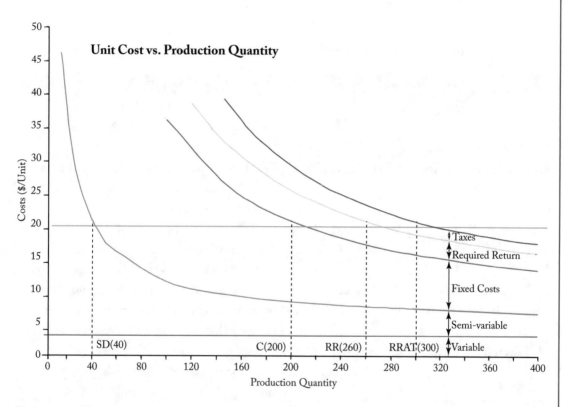

Figure 4.2: Unit costs vs. production quantity for the production-based model [4, 5].

Table 4.2: Shutdown points and actions/implications for production quantity-based model [4]

Production Level Range	Action/Implication
1. Zero to Shutdown Level (SD) (0–40 units)	Do not accept order as all of the out-of-pocket costs (variable and semi-variable costs) will not be recovered.
2. Shutdown Level (SD) to Cost Level (C) (40–200 units)	Will recover the out-of-pocket costs, but not all of the fixed costs. Accept only if no better opportunities are available.
3. Cost Level (C) to Required Return Level (RR) (200–260 units)	Will recover all costs, but will not obtain the desired level of required return. Accept if no better opportunities are available.
4. Required Return Level (RR) Level to Required Return After Taxes Level (RRAT) (260–300 units)	Have succeeded in making the required return on a pre-tax basis, but not on an after-tax basis. Accept unless better opportunities are available.
5. Greater than Required Return After Taxes Level (RRAT) (>300 units)	Will recover all costs and exceed required return on an after-tax basis. Accept as this is usually a rare and highly profitable event.

4.5   PRODUCTION TIME-BASED BREAKEVEN EXAMPLE

The production time-based method is newer and is used as overheads are becoming a major component of costs and is typically based upon time. All four breakeven points will be illustrated and this is the same example used at the ASEE [4] conference. The data for the production quantity-based model is listed in Table 4.3.

Table 4.3: Data for production time-based breakeven analysis [1]

Item	$/Hour	$	Decimal
Sales Revenue		13,000	
Production (Manufacturing) Costs			
Fixed Costs		2,000	
Variable Costs	18		
Semi-variable Costs	2	1,000	
Overhead Costs	20		
Required Return (Profit)	10		
Tax Rate (40%)			0.40

Notice that the overhead costs are a separate item whereas they were included in the production costs in the quantity-based model. The required return can be used as an hourly rate as illustrated in this example, but it can also be listed as a specific amount.

Let y = production hours and then the breakeven points can be calculated as follows.

(a) Shutdown Breakeven Point (SD)

$$Revenues = Production\ Costs$$
$$Revenues = Fixed\ Costs + Variable\ Costs + Semi\text{-}variable\ Costs$$
$$13,000 = 2,000 + 18y + 2y + 1,000$$
$$13,000 = 3,000 + 20y$$
$$20y = 10,000$$
$$y = 500\ hours$$

(b) Cost Breakeven Point (C)

$$\text{Revenues} = \text{Total Costs}$$
$$\text{Revenues} = \text{Production Costs} + \text{Overhead Costs}$$
$$13,000 = 3,000 + 20y + 20y$$
$$13,000 = 3,000 + 40y$$
$$40y = 10,000$$
$$y = 250 \text{ hours}$$

(c) Required Return Breakeven Point (RR)

$$\text{Revenues} = \text{Total Costs} + \text{Required Return}$$
$$\text{Revenues} = 3,000 + 40y + 10y$$
$$13,000 = 3,000 + 50y$$
$$50y = 10,000$$
$$y = 200 \text{ hours}$$

(d) Required Return After Taxes Breakeven Point (RRAT)

$$\text{Revenues} = \text{Total Costs} + \text{Required Return After Taxes} + \text{Taxes on Total Required Return}$$
$$\text{Revenues} = \text{Total Costs} + \text{Required Return After Taxes}/(1 - \text{Tax Rate})$$
$$\text{Revenues} = 3,000 + 40y + 10y/(1.0 - 0.40)$$
$$13,000 = 3,000 + 40y + 16.66y$$
$$56.66y = 10,000$$
$$y \approx 176 \text{ hours } (176.47)$$

Note that to obtain a required return of $10y$ after taxes one must earn $16.66y$ as the 40% taxes would be $6.66y$. To obtain the pre-tax required return, one can use the expression:

$$\text{Required Return Including Taxes} = \text{Required Return After Taxes}/(1.0 - \text{Tax Rate}). \quad (4.5)$$

The tax rate is expressed as a decimal.

The results and actions for the various breakeven points are presented in Table 4.4. The Total Cost vs. Production Times in Figure 4.3 shows the breakeven points in a manner similar to the production unit costs where the revenue is a horizontal line. The profitability curve in Figure 4.4 highlights that lower production times have a major impact upon profitability.

The primary advantage of the profitability plot is that it easily shows the importance of doing things faster and how that improves profitability. This also can be used to estimate the cost of delay upon profitability such as a machine having down-time, delivery delay, weather delay, etc.

Table 4.4: Shutdown points and actions/implications for the production time-based model [1, 5]

Production Level Range	Action/Implication
1. Shutdown Level (SD) or higher (> 500 production hours)	Do not accept order as all of the direct production costs will not be recovered and none of the overhead or return will be recovered.
2. Breakeven Cost Level (C) to Shutdown Level (SD) (250–500 production hours)	Will recover all of the production costs and some of the overhead costs. None of the required return will be recovered.
3. Required Return Level (RR) to Breakeven Cost Level (C) (200–250 production hours)	Will recover all costs, but will not obtain the desired level of required return before taxes. Accept unless better opportunities are available.
4. Required Return After Taxes Level (RRAT) to Required Return Level (RR) (176–200 production hours)	Have succeeded in making the required return on a pre-tax basis, but not on an after-tax basis. Accept unless better opportunities are available.
5. Less than Required Return After Taxes Level (RRAT) (< 176 production hours)	Will recover all costs and exceed required return on an after-tax basis. Accept as this is usually a highly profitable rare event.

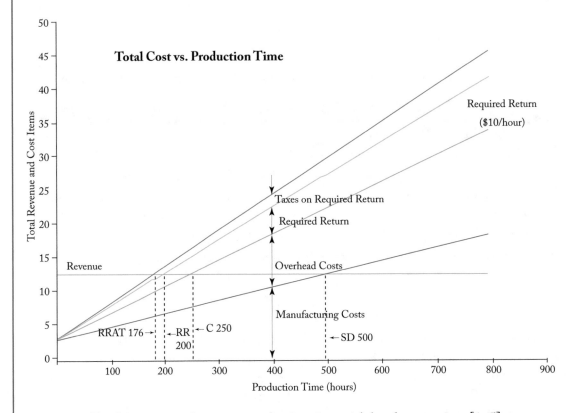

Figure 4.3: Total revenues and costs vs. production time with breakeven points [1, 5].

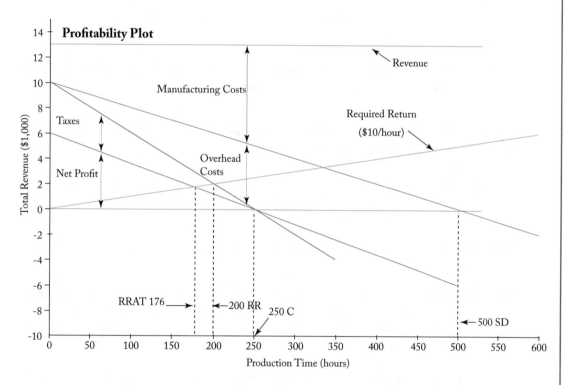

Figure 4.4: Profitability vs. production time with breakeven points [1, 5].

4.6 SUMMARY

The traditional breakeven charts usually have only one breakeven point at cost, but three other points can be determined and they are the shut-down point, the breakeven point at required return, and the breakeven point at required return after taxes. The time-based system is being used for allocating overheads and can also be used for determining the four breakeven point using hour units. The plot of total revenue and the various costs vs. production illustrates the breakeven points on the production quantity basis, and the profitability plot of profitability vs. production hours illustrates the breakeven points on the production time basis.

4.7 REFERENCES

[1] Creese, Robert C., Time-based breakeven analysis and costing, *AACE International Transactions*, ABC.02, AACE International, Morgantown WV, pp. ABC.02.1–ABC.02.6, 1998. 43, 46, 51, 53, 54, 55

Reprinted with the permission of AACE International, 1265 Suncrest Towne Centre Dr., Morgantown, WV, 26505, Phone 304-296-8444.
Internet: http://web.aacei.org

[2] Creese, Robert C., AACE.02, Time-based breakeven analysis, *Joint Cost Management Societies Proceedings*, pp. AACE 02.01–AACE 02.07, 1998.

[3] Creese, Robert C., A new breakeven analysis uses production versus quantity, *Modern Castings*, pp. 52–53, March 1996.

[4] Creese, Robert C., Time-based versus quantity based breakeven analysis, *Proc. of the American Society for Education Annual Conference and Exposition*, pp. 9.1308.1–9.1308.17, 2004. Selected items are Reprinted with permission of American Society for Engineering Education. 45, 46, 48, 49, 50, 51

[5] Creese, Robert and Thiruvalam, Kedhar P., Power Point Presentation, "Breakeven Analysis," prepared in 2009 for presentation at Metal Casting Seminar. 43, 48, 49, 53, 54, 55

[6] Dieter, George E., Costs and related aspects of materials selection, *ASM Handbook Volume 20 Materials Selection and Design*, ASM International, Metals Park, OH, pp. 248–249. 43

4.8 EVALUATIVE QUESTIONS

1. When plotting the production breakeven charts the fixed costs traditionally were plotted first.

(a) What happens when that is done on the Total Cost vs. Production Quantity graph?

(b) What happens if the variable cost is plotted first on the Unit Cost vs. Production Quantity graph?

2. What are the four breakeven points if the variable costs are 8 $/unit instead of 3 $/unit in Table 4.1?

3. Plot both the total cost vs. production quantity breakeven chart similar to Figure 4.1 and the unit cost breakeven vs. production quantity chart similar to Figure 4.2 when the variable costs are 8 $/unit.

4. What are the four breakeven points if the Required Return in Table 4.1 was $1,200 instead of $900?

5. What are the four breakeven points if the semi-variable costs in Table 4.1 were $7\times + 900$ instead of $2\times + 600$?

6. What are the four breakeven points for the Production Time-Based Breakeven analysis if the fixed costs in Table 4.3 were $5,000 instead of $2,000?

7. Make the Total Revenues and Costs vs. Production Time results similar to Figure 4.3 and the Profitability vs. Production Time results similar to Figure 4.4 showing the four breakeven points on each of the graphs.

8. The data for a production quantity-based breakeven problem is in Table 4.5. Calculate the following items:

(a) Shutdown breakeven point-units

(b) Cost breakeven point-units

(c) Required return breakeven point-units

(d) Required return after taxes breakeven point

(e) Draw the Total Revenue and Costs vs. the Production Quantity showing the four breakeven points.

9. The data for a production time-based breakeven problem is in Table 4.6. Calculate the following items:

(a) Shutdown breakeven point-units

(b) Cost breakeven point-units

(c) Required return breakeven point-units

(d) Required return after taxes breakeven point

(e) Draw the Profitability Plot showing the four breakeven points.

Table 4.5: Data for production quantity-based breakeven analysis

Item	$/Unit	$	Decimal
Sales Revenue	20		
Production (Manufacturing) Costs			
Fixed Costs		3,600	
Variable Costs	3		
Semi-variable Costs	2	900	
Required Return (Profit)		1,200	
Tax Rate (20%)			0.20

Table 4.6: Data for production time-based breakeven analysis

Item	$/Hour	$	Decimal
Sales Revenue		13,000	
Production (Manufacturing) Costs			
Fixed Costs	10		
Variable Costs	6		
Semi-variable Costs	2	3,000	
Overhead Costs	10		
Required Return (Profit)		1,200	
Tax Rate (20%)			0.20

CHAPTER 5

Earned Value Management

5.1 INTRODUCTION

Earned Value Management (EVM) is a tool used to measure project performance with respect to time and cost. It is typically used to measure performance in project work which involves a single highly complex item whereas breakeven analysis is used more frequently in manufacturing and other areas where the output involves a large quantity of a single product or closely related products. EVM focuses on schedule or time, and costs. EVM can be applied to any project such as developing new products, research projects, equipment installation projects, and any project that involves a schedule and costs. Typical EVM projects are large construction projects buildings, highways, bridges, or dams and other very large projects such as aircraft carriers, missions to the moon or Mars, and the development of innovative new products such as the i-Phone or self-driving vehicles. EVM can indicate performance problems such as cost overruns and/or schedule delays during the project to warn management to take action so the project can achieve a successful completion. When the project is large, EVM is used for measuring progress in terms of cost and schedule to specific milestones of the project. Two sources [1, 2] of the fundamental concepts of EVM are used in the development of the following relationships and equations. The data used is from earlier examples from AACE International [1], which has become a base model for comparing methods for evaluating the traditional EVM approach and the newer time-based EVM approach developed by Lipke [3, 4]. The two approaches will be illustrated to show the differences as the EVM method shows the schedule variance in terms of hours or dollars and the time-based method calculates the schedule variance in terms of time periods rather than dollars. The differences can become large at the end of the project when projects have significant delays or early completion.

5.2 EARNED VALUE MANAGEMENT PERFORMANCE PARAMETERS

Three elements are used to measure project performances in EVM:

1. the Planned Value (PV) of the work scheduled or the Budgeted Cost of Work Scheduled (BCWS);

2. the Earned Value (EV) of the work accomplished or the Budgeted Cost of Work Performed (BCWP); and

3. the Actual Cost (AC) of the work accomplished or the Actual Cost of Work Performed (ACWP).

The values are traditionally measured in monetary units, such as dollars, or measured in work hours, and sometimes both are used. Work hours are often used at the work site and converted to monetary units at the management level. The earned value is used in both the schedule performance and in the cost performance measurements.

The traditional performance measures derived from these elements are the Schedule Variance (SV), the Cost Variance (CV), the Schedule Performance Index (SPI), and the Cost Performance Index (CPI), which are presented in the following equation forms. The schedule values will be presented first:

$$SV = EV - PV, \tag{5.1}$$

where

SV = schedule variance
EV = earned value
PV = planned value.

If the schedule variance is positive, the project is ahead of schedule whereas if the schedule variance is negative, the project is behind schedule. This same information can be obtained from the schedule performance index where:

$$SPI = EV/PV, \tag{5.2}$$

where

SPI = schedule performance index
EV = earned value
PV = planned value.

A schedule performance index of 1.0 indicates the project is on schedule, a schedule performance index greater than 1.0 indicates the project is ahead of schedule, and a schedule performance index less than 1.0 indicates the project is behind schedule.

The cost performance equations are similar to the schedule performance equations, but use the actual cost and earned value. The equations are:

$$CV = EV - AC, \tag{5.3}$$

where

CV = cost variance
EV = earned value
AC = actual cost.

If the cost variance is positive, the project is under budget whereas if the cost variance is negative, the project is over budget. This same information can be obtained from the cost performance index where:

$$CPI = EV/AC, \qquad (5.4)$$

where

> CPI = cost performance index
> EV = earned value
> AC = actual cost.

A cost performance index of 1.0 indicates the project is on budget, a cost performance index greater than 1.0 indicates the project is under budget, and a cost performance index less than 1.0 indicates the project is over budget.

These values can be kept on a periodic basis, such as weekly, and on a cumulative basis for monitoring the project performance and direction and can give indications as to what corrective actions need to be taken. Note that Earned Value (EV) is used in both the cost and schedule indices and variances.

Another item for consideration is to estimate the completion of the project. The completions can be calculated for the cost completion and the schedule completion. There are various methods, but the one most often used is to consider the work cost in the future will be at the planned rate and this results in:

$$EAC(c) = PV_{ct} + (AC - EV), \qquad (5.5)$$

where

> $EAC(c)$ = estimate at completion-cost
> AC = actual cost to date
> EV = earned value to date
> PV_{ct} = planned value for project completion cost.

The completion can also be predicted with respect to time as well, using the planned value and is expressed as:

$$EAC(t) = PV_{ct} + (PV - EV), \qquad (5.6)$$

where

> $EAC(t)$ = estimate at completion (time)
> PV = planned value to date
> EV = earned value to date
> PV_{ct} = planned value for project completion time.

5.3   EXAMPLE PROBLEM USING TRADITIONAL EARNED VALUE MANAGEMENT

An example problem, adopted from the AACEI Skills and Knowledge of Cost Engineering, was expanded and is presented on a project with a delay of one week over the life of the project. The information is presented in Table 5.1. The tables and figures in this chapter are from the work of Yi Fang which are published in her thesis [5] and also in *Cost Engineering* [6] and are published with permission of AACE International. To illustrate how the values are calculated, Week 3 will be analyzed in detail for the variances and indexes. The planned time for the project total completion or PV_{ct} is 440 hours. For the period values for week three using Equations (5.1) to (5.4), one obtains:

$$SV(3) = EV - PV = 65 - 45 = +20$$
$$SPI(3) = EV/PV = 65/45 = 1.44 > 1.00$$
$$CV(3) = EV - AC = 65 - 62 = +3.0$$
$$CPI(3) = EV/AC = 65/62 = 1.05 > 1.00.$$

This indicates that the schedule and cost performance indices were both greater than 1.00 for Week 3 and thus on time and under budget for the week.

For the cumulative values, the symbol SV_{c3} implies the schedule variance for Week 3. The cumulative values for week three are:

$$SV_{c3} = EV_{c3} - PV_{c3} = 110 - 120 = -10$$
$$SPI_{c3} = EV_{c3}/PV_{c3} = 110/120 = 0.92 < 1.0$$
$$CV_{c3} = EV_{c3} - AC_{c3} = 110 - 109 = +1.0$$
$$CPI_{c3} = EV_{c3}/AC_{c3} = 110/109 = 1.01 > 1.0$$

$$EAC(c) = PV_{ct} - CV_{c3} = PV_{ct} - (EV_{c3} - AC_{c3}) = 440 - 1$$
$$= 439 = 440 - (110 - 109) = 439$$
$$\text{Percent Completion} = (EV_{c3}/PV_{ct}) \times 100 = (110/440) \times 100 = 25\%$$
$$EAC(t) = PV_{ct} - SV_{c3} = PV_{ct} - (EV_{c3} - PV_{c3}) = 440 - (-10)$$
$$= 450 + (120 - 110) = 450.$$

This indicates that for the cumulative first three weeks, the project is behind schedule, slightly under budget, and expected to be completed at 450 hours based upon schedule performance (typically used) or at 439 hours based upon cost performance. The project is 25% complete at the end of the third week. If the percent completion is known, it can be used to determine the earned value. That is:

$$\text{Earned Value} = (\text{Percent Completion as decimal}) \times \text{Planned Value.} \tag{5.7}$$

The cumulative *PV*, *AC*, and *EV* values are plotted in Figure 5.1. Note that the period values and cumulative values of the *SPI* and *CPI* can be quite different, and indicated by the schedule parameters in Table 5.1. One problem with the schedule indices is that the *SV* tends to 0 and the *SPI* goes to 1.0 regardless of whether or not the project is delayed. This occurs as the planned and earned values will be equal when the project is completed, regardless if it is early, on time, or late. This has led to a new calculation procedure for the Schedule Variance and the Schedule Performance Index on a time based analysis rather than on a cost-based (or work hour-based) analysis.

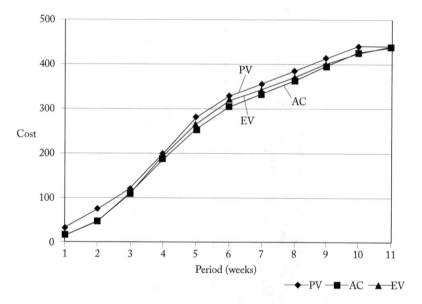

Figure 5.1: Planned value, earned value, and actual cost [6].

5.4 EXAMPLE PROBLEM USING EARNED SCHEDULE IN EARNED VALUE MANAGEMENT

Lipke [3, 4] introduced the concept of Earned Schedule (*ES$_c$*) which is similar to the Earned Value, but Earned Schedule is in time units rather than hours or dollars. Fang [5, 6] utilized the concept and applied it to the AACE International data [1] in her M.S. thesis. The earned schedule values are used for cumulative values, not period values. The following calculations are utilizing the data in Table 5.2 which also contains the cumulative values of Table 5.1.

The cumulative time values has two PV values for *N* and *N* + 1. The *AT* value is usually greater than *N*, but if a project finishes ahead of schedule *N* can equal or be greater than *AT*.

Table 5.1: Period and cumulative data for earned value approach for delayed project [5, 6]

Period (Week) (AT)	Period Values							Cumulative Values						
	PV (BCWS)	AC (ACWP)	EV (BCWP)	SV	SPI	CV	CPI	PVc (BCWS)	ACc (ACWP)	EVc (BCWP)	SPIc	SVc	CPIc	CVc
1	30	16	15	-15	0.50	-1.0	0.94	30	16	15	0.50	-15	0.94	-1.0
2	45	31	30	-15	0.67	-1.0	0.97	75	47	45	0.60	-30	0.96	-2.0
3	45	62	65	+20	1.44	+3.0	1.05	120	109	110	0.92	-10	1.01	+1.0
4	80	78	85	+5	1.06	+7.0	1.09	200	187	195	0.98	-5	1.04	+8.0
5	80	66	70	-10	0.88	+4.0	1.06	280	253	265	0.95	-15	1.05	+12.0
6	50	51	55	+5	1.10	+4.0	1.08	330	304	320	0.97	-10	1.05	+16.0
7	25	30	25	0	1.00	-5.0	0.83	355	334	345	0.97	-10	1.03	+11.0
8	30	30	25	-5	0.83	-5.0	0.83	385	364	370	0.96	-15	1.02	+6.0
9	30	33	30	0	1.00	-3.0	0.91	415	397	400	0.96	-15	1.01	+3.0
10	25	28	25	0	1.00	-3.0	0.89	440	425	425	0.97	-15	1.00	0.0
11	0	14	15	+15	N/A	+1.0	1.07	440	439	440	1.00	0	1.00	+1.0

The equation for ES_c is mathematically expressed as:

$$ES_c = N + (EV_c - PV_c(N))/(PV_c(N+1) - PV_c(N)), \tag{5.8}$$

where

ES_c = cumulative earned schedule at time period N
N = period where the cumulative PV is less than the current cumulative EV
$EV_c(AT)$ = cumulative earned value at actual time, AT
$PV_c(N)$ = cumulative planned value at period N
$PV_c(N+1)$ = cumulative planned value at next period, $N+1$
AT = actual time or current time period.

For Period 3 (that is where $AT = 3$ and $N = 2$),

$$ES_3 = 2 + (110 - 75)/(120 - 75) = 2 + 0.777 = 2.78 \text{ periods (weeks)}.$$

If, at the start of the project, the PV is greater than the EV, then the ES will be less than one and can be calculated by

$$ES_c = (PV_1 - EV_1)/PV_1. \tag{5.9}$$

Thus, for Period 1, where $AT = 1$

$$ES_1 = (30 - 15)/30 = 0.5 \text{ period (week)}.$$

This calculation can be repeated for the initial periods until ES_c exceeds PV_1 and then Equation (5.9) can be used and this is when $N = 1$. The schedule variance, based upon time, is calculated by:

$$SV(AT) = ES - AT, \tag{5.10}$$

where

$SV(AT)$ = schedule variance based upon time AT
ES = earned schedule
AT = actual time or current time period
$SV(3) = 2.78 - 3.00 = -0.22 \text{ periods (weeks)}$

which implies the project is 0.22 time periods behind schedule at the end of time period 3.

The schedule performance index based upon time is:

$$SPI(AT) = ES/AT. \tag{5.11}$$

If one examines Period 3, then

$$SPI(3) = 2.78/3.0 = 0.93.$$

The first cumulative value of PV below EV is when $PV = 30$ and $EV = 45$ which occurs in period 2, thus

$$AT = 2$$
$$EV_C = 45$$
$$N = 1$$
$$PV_c = 30$$
$$PV_c(N + 1) = 75$$

and

$$ES_c = N + (EV_c - PV_c(N))/(PV_c(N + 1) - PV_c(N))$$
$$ES(2) = 1 + (45 - 30)/(75 - 30) = 1 + 0.3333 = 1.33 \text{ periods (weeks)}.$$

Thus, the time-based schedule variance using Equation (5.10) is:

$$SV(2) = 1.33 - 2.00 = -0.67 \text{ periods (weeks)}.$$

This implies that the project is 0.67 time periods behind schedule. The schedule performance index based upon time using Equation (5.2) is:

$$SPI(2) = ES(2)/AT = 1.33/2.00 = 0665 = 0.67.$$

Now examining the data for the last period, that is period 11, note that

$$AT = 11$$
$$EV_{(AT)} = 440.$$

The first cumulative value of PV below $EV_{(11)}$ is 415 which occurs in period $N = 9$, thus

$$N = 9$$
$$PV_C = 415$$
$$PV_C(N + 1) = 440$$

and

$$ES_c = N + (EV_c - PV_c(N))/(PV_c(N + 1) - PV_c(N))$$
$$ES = 9 + (440 - 415)/(440 - 415) = 9 + 1 = 10 \text{ periods (weeks)}.$$

Thus, the time-based schedule variance using Equation (5.10) is:

$$SV(AT) = ES - AT$$
$$SV(11) = 10 - 11 = -1 \text{ periods (weeks)}.$$

This implies that the project is 1.0 time period behind schedule. The final schedule variance is the difference in time values (AT) when the cumulative planed value (PV_c) reaches the project completion time minus the (AT) when the earned value ($EV(N + 1)$) first reaches the project complete value. The schedule performance index based upon time using Equation (5.11) is:

$$SPI(11) = ES/AT = 10/11 = 0.909 = 0.91.$$

The time-based values are the actual values, that is a schedule variance of -1.0 and schedule performance index of 0.909 instead of the traditional schedule variance of 0 and SPI of 1.0. Table 5.2 indicates the cumulative-based and cumulative time-based values. The major differences are in the schedule variance and the schedule performance index. The schedule variance is based upon dollars or work hours, whereas the earned schedule is based upon amount of work completed for that particular time period.

Figure 5.2 show the comparison of the SPI values based upon the cumulative dollar (or work-hour) base ($SPI(\$)$) vs. the SPI values based upon the cumulative project time ($SPI(t)$) over the project life for a delayed project. There would also be a difference if the project was completed early. The differences are a result of the EVM work hour approach must end with a $SPI(\$)$ at 1.0, regardless of the actual completion time whereas $SPI(AT)$ is at 0.91.

The major differences between the two SPI indices occur at the end of the project life. The initial differences are primarily due to the difference in units; that is dollars vs. the time estimate calculation. The time-based calculation is the relative amount of work earned for the period vs. the amount planned for the period. Table 5.3 is a combination of Tables 5.1 and 5.2 to indicate all the information in a single table.

5.5 SUMMARY

The parameters used to measure project performance are the Planned Value, the Earned Value, and the Actual Cost of the work performed. The performance measures evaluated were the schedule variance, the schedule performance index, the cost variance, the cost performance index, the time estimated at completion, the time-based schedule variance, and the time-based schedule performance index. Indices which are greater than unity are favorable (under budget or ahead of schedule) and those less than unity are unfavorable (over budget or behind schedule). The traditional schedule variance and schedule performance index calculations give less accurate results near the project completion than the time based schedule variance and schedule performance index. These performance measures are critical for long-term projects success so adjustments can be made to improve project performance during the project execution and to minimize project delays and cost overruns.

Table 5.2: Time-based calculations for earned schedule [5, 6]

Period (Week) (AT)	Earned Value Cumulative Values							Earned Schedule Cumulative Time Values						
	PV (BCWS)	AC (ACWP)	EV (BCWP)	SV_c	SPI_c	CV_c	CPI_c	N	EV (AT)	PV (N+1)	PV (N)	ES	SV (AT)	SPI (AT)
1	30	16	15	-15	0.5	-1	0.94	N/A	N/A	N/A	N/A	0.5	-0.5	0.5
2	75	47	45	-30	0.6	-2	0.96	1	45	75	30	1.33	-0.67	0.67
3	120	109	110	-10	0.92	1	1.01	2	110	120	75	2.78	-0.22	0.93
4	200	187	195	-5	0.98	8	1.04	3	195	200	120	3.94	-0.06	0.98
5	280	253	265	-15	0.95	12	1.05	4	265	280	200	4.81	-0.19	0.96
6	330	304	320	-10	0.97	16	1.05	5	320	330	280	5.80	-0.20	0.97
7	355	334	345	-10	0.97	11	1.03	6	345	355	330	6.60	-0.40	0.94
8	385	364	370	-15	0.96	6	1.02	7	370	385	355	7.50	-0.50	0.94
9	415	397	400	-15	0.96	3	1.01	8	400	415	385	8.50	-0.50	0.94
10	440	425	425	-15	0.97	0	1.0	9	425	440	415	9.40	-0.60	0.94
11	440	439	440	0	1.0	1	1.0	10	440		440	10.0	-1.00	0.91

Table 5.3: Period values, cumulative values, and time-based values [5, 6]

Period (Week) (AT)	Period Values							Cumulative Values							Cumulative Time-Based Values						
	PV (BCWS)	AC (ACWP)	EV (BCWP)	SV	SPI	CV	CPI	PVc (BCWS)	ACc (ACWP)	EVc (BCWP)	SPIc	SVc	CPIc	CVc	N	EV (AT)	PV (N)	PV (N+1)	ES (AT)	SV (AT)	SPI (AT)
1	30	16	15	-15	0.50	-1.0	0.94	30	16	15	0.50	-15	0.94	-1.0	NA	NA	NA	NA	0.5	-0.5	0.5
2	45	31	30	-15	0.67	-1.0	0.97	75	47	45	0.60	-30	0.96	-2.0	1	45	30	75	1.33	-0.67	0.67
3	45	62	65	+20	1.44	+3.0	1.05	120	109	110	0.92	-10	1.01	+1.0	2	110	75	120	2.78	-0.22	093
4	80	78	85	+5	1.06	+7.0	1.09	200	187	195	0.98	-5	1.04	+8.0	3	195	120	200	3.94	-0.06	0.98
5	80	66	70	-10	0.88	+4.0	1.06	280	253	265	0.95	-15	1.05	+12.0	4	265	200	280	4.81	-0.19	0.96
6	50	51	55	+5	1.10	+4.0	1.08	330	304	320	0.97	-10	1.05	+16.0	5	320	280	330	5.8	-0.2	0.97
7	25	30	25	0	1.00	-5.0	0.83	355	334	345	0.97	-10	1.03	+11.0	6	345	330	355	6.6	-0.4	0.94
8	30	30	25	-5	0.83	-5.0	0.83	385	364	370	0.96	-15	1.02	+6.0	7	370	355	385	7.5	-0.5	0.94
9	30	33	30	0	1.00	-3.0	0.91	415	397	400	0.96	-15	1.01	+3.0	8	400	385	415	8.5	-0.5	0.94
10	25	28	25	0	1.00	-3.0	0.89	440	425	425	0.97	-15	1.00	0.0	9	425	415	440	9.4	-0.6	0.94
11	0	14	15	+15	N/A	+1.0	1.07	440	439	440	1.00	0	1.00	+1.0	10	440	440	440	10	-1	0.91

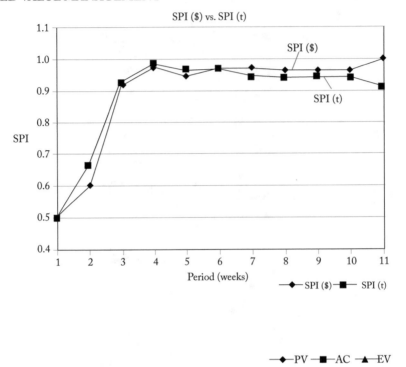

Figure 5.2: Comparison of cost-based *SPI($)* and time-based *SPI(t)* [5, 6].

5.6   REFERENCES

[1] AACE International, *Skills and Knowledge of Cost Engineering*, 5th ed., Chapters 14, 15, 2004. 59, 63

[2] Creese, R. C. and Adithan, M., Earned value management concepts, *Strategic Cost Analysis*, New Academic Science Limited, Tunbridge Wells, Kent, UK, pp. 13–23, 2012. 59

[3] Lipke, W., A study of the normality of earned value management indicators, *Measurable News*, pp. 1–16, 2002. 59, 63

[4] Lipke, W., Schedule is different, *Measurable News*, March 2003. 59, 63

[5] Fang, Y., *Estimate at Completion for Construction Projects*, Master's Problem Report, Industrial and Management Systems Engineering, West Virginia University, 2008. 62, 63, 64, 68, 69, 70

[6] Creese, R. C. and Fang, Yi, Time-based schedule performance index, *Cost Engineering*, Vol. 53, No. 3, pp. 18–20, March 2010. Reprinted with Permission of AACE Interna-

tional, 1265 Suncrest Towne Centre Drive, Morgantown, WV 26505 Phone (304) 296-8444. `http;//web.aacei.org` 62, 63, 64, 68, 69, 70

5.7   EVALUATIVE QUESTIONS

1. Angel Construction Company was awarded a contract for $2,000,000 to build a cathedral with a budget of 200,000 hours (PV_c). Management wants a progress update and the end of week six and the following data had a budget of 200,000 hours (PV_c). Management wants a progress update at the end of week six and the following data was obtained.

   Planned Hours   80,000

   Actual Hours Spent   75,000

   Percent Completion (based on Earned Hours)   35%

   (a) How many hours has the project earned?

   (b) What is the cost variance (hrs)?

   (c) What is the schedule variance (hrs)?

   (d) What is the CPI?

   (e) What is the SPI?

   (f) What is the estimated time for completion?

   (g) What is the estimated cost at completion?

2. Pumpkin Research and Development was awarded a contract for $25,000,000 to build a machine to use supersonic sound waves to remove body fat at a rate of 5 kg/hour. The budget is estimated to take 50,000 hours (PV_c). Management wants a progress update and the end of week six and the following data was obtained.

   Planned Hours   7,000

   Actual Hours Spent   8,000

   Percent Completion (based on Earned Hours)   20%

   (a) How many hours has the project earned?

   (b) What is the cost variance (hrs)?

   (c) What is the schedule variance (hrs)?

   (d) What is the CPI?

   (e) What is the SPI?

   (f) What is the estimated time for completion?

   (g) What is the estimated cost at completion?

3. Use the data of PV, AC, and EV for time periods 6 and 7 in Table 5.1 and calculate the values for SV, SPI, CV, and CPI for the period and cumulative values to three decimals (for non-integer terms) and compare the values with those given in Table 5.1.

4. Calculate the earned value cumulative values of SV_c, SPI_c, CV_c, CPI_c, and earned schedule cumulative time values of $ES, SV(AT)$, and $SPI(AT)$ to three decimals (for non-integer terms) for weeks 6 and 7 and compare the values with those in Table 5.2 for weeks 6 and 7.

5. Change the PV values for Periods 6, 7, and 8 from 50, 25, 30 to 45, 35, 25 and note the differences in the $SV, SPI, CV, CPI, SPI_C, SV_C, CPI_C, CV_C, ES, SV(AT)$, and $SPI(AT)$ values from those in Table 5.3 for those three periods.

6. Use the data in Table 5.4 and calculate the period and cumulative values of SV, SPI, CV, CPI, and the time-based values of $SV(t)$ and $SPI(t)$ for all the periods for the project manager of the Tibet Construction Company. Plot the $SPI(\$)$ and $SPI(t)$ values for the project over the five time periods. Discuss the differences between the SV_c and $SV(AT)$ and the SPI_c and $SPI(AT)$.

Table 5.4: Weekly data of the Tibet Construction Company on the West China windmill construction project

Period (Week) (AT)	Period Values							Cumulative Values								Cumulative Time-Based Values					
	PV (BCWS)	AC (ACWP)	EV (BCWP)	SV	SPI	CV	CPI	PVc (BCWS)	ACc (ACWP)	EVc (BCWP)	SVc	SPIc	CVc	CPIc	N	PV (N)	PV (N+1)	EV (AT)	ES (AT)	SV (AT)	SPI (AT)
1	40	25	20																		
2	40	35	35																		
3	40	50	45																		
4	40	40	40																		
5	0	25	20																		

PART II

Tools for Economic Evaluations

CHAPTER 6

Fundamental Definitions, Terms, and Concepts for Technical Economic Evaluations

6.1 INTRODUCTION

The previous chapters have focused on macro-concepts such as financial statements, profits and cash flows, the Purcell Diagram, breakeven analysis, ABC and time-based evaluations, estimating ranges, and accuracies. Now the micro-concepts need to be presented in detail so that the expressions developed and methods applied using these items in the following chapters will be better understood. The primary focus of this chapter will be on interest, the various types of interest, inflation, constant and current currency, and exchange rates. This material is available in many references on engineering economy and much of this is based upon materials developed for short courses given in the past [1–3] and a book [4] published based on the materials in these short courses.

6.2 FUNDAMENTAL TERMS RELATED TO INTEREST CALCULATIONS

6.2.1 INTEREST AND INTEREST RATE

There are many types of interest and two primary definitions of interest are the rate charged for the investment of capital and the return rate for the investment of capital.

(1) The cost for the use of capital which is also referred to as the time value of money. (This is the view of the borrower who considers it as a cost or rate for the use of the capital borrowed.)

(2) The monetary return or rate of return which is necessary to divert money into long-term investments. (This is the view of the lender who considers it as a rate of return on the investment.)

The interest rate is the ratio of the interest amount accrued in the time period to the amount owed at the start of that period. There are two major types of interest commonly used and they are simple interest and compound interest.

Simple Interest

Simple interest is the interest rate that determines the interest amount only on the principal amount. Simple interest can be defined as follows.

(1) Interest charges are only charged on the principal at the start of the period and not on any additions or deletions made during the period.

(2) Interest is calculated only on the investment at the end of the period, but it is not included as part of the investment for the following periods.

The interest calculations for simple interest problems are presented by first calculating the total amount of interest and then the total amount due at the end of the period which includes the principal. The total interest amount is:

$$I = niP, \tag{6.1}$$

where

I = total interest amount due
i = interest rate per unit time period (frequently a year)
n = number of time periods
P = principal amount or initial investment amount at the beginning of the period.

An amount P will be invested for n time periods at an interest rate i and the amount due at the end of the n periods will be the Principal, (P), and the total amount of Interest (I). This is usually referred to as the Future Worth (F) amount or Future worth, thus:

$$F = P + I = P + niP = P(1 + in). \tag{6.2}$$

To illustrate the application of the formula, let

$P = \$500$
$n = 3$
$i = 10\%$ or as a decimal, 0.10.

Thus, the amount paid at the end of three years is calculated as:

$$F = 500(1 + 0.10 \times 3) = 500(1.3) = \$650$$

= \$500	+ \$50	+ \$50	+ \$50
Principal	Interest for First Year	Interest for Second Year	Interest for Third Year.

Note that the principal does not change over the n periods and no interest is earned on the accumulated interest.

Compound Interest

Compound interest is the interest which is most commonly applied. Compound interest is interest on the principal amount and the interest on the previous amounts of interest earned. Two definitions [4] of compound interest are as follows.

(1) The type of interest that is periodically added to the amount of principal, investment, or loan so that the subsequent interest rate is based on the cumulative amount of principal plus total interest.

(2) The type of interest that is charged on any previous interest earned in any time period as well as on the principal.

When considering compound interest calculations, there are two approaches of calculating interest.

(1) Discrete Compound Interest Rate is the interest rate is applied at the end of each time period and amount determined is considered only in the following time periods. This is the most common form of compound interest rate application.

(2) Continuous Compound Interest Rate is the interest rate is applied continuously during each time period. This is less commonly applied today, but with businesses being open 24-7, it may be applied more in the future. It is often applied on certificates of deposit.

An example of each approach will be presented, starting with the discrete compound rate applications. The following table will illustrate the approach for determining the Future Amount (F) at the end of n periods starting with the initial Principal Amount (P).

Time Period	Initial Amount (Beginning of Period)	+	Interest Amount	=	Total Amount (End of Period)
1	P	+	iP	=	$P(1+i)$
2	$P(1+i)$	+	$iP(1+i)$	=	$P(1+i)^2$
3	$P(1+i)^2$	+	$iP(1+i)^2$	=	$P(1+i)^3$
4	$P(1+i)^3$	+	$iP(1+i)^3$	=	$P(1+i)^4$

and in general for any value of n

n	$P(1+i)^{(n-1)}$	+	$iP(1+i)^{(n-1)}$	=	$P(1+i)^n = F$

Thus, the expression for discrete compound interest would be:

$$F = P(1+i)^n. \tag{6.3}$$

Using the data of the previous simple interest example and now applying the discrete compounding interest formula, let:

$P = \$500$

$n = 3$

$i = 10\%$ or as a decimal, 0.10

Time Period	Initial Amount (Beginning of Period)	Interest Amount	Total Amount (End of Period)
1	500	50	550
2	550	55	605
3	605	60.50	665.50.

This can be determined by Equation (6.3) as:

$$F = P(1 + i)^n = 500(1 + 0.10)^3 = \$665.50$$

Note that the total amount is $665.50 for discrete compounding vs. $650. for simple interest. The difference in total interest or return is $15.50 and which is approximately a 2% difference over the 3-year period.

Continuous compounding is compounding throughout the period and not only at the end of the period. The expression is obtained by letting the number of compounding periods go to infinity and thus the interest is considered as the effective interest over the entire period and r is the nominal interest per year:

$$i_{eff} = \lim_{m \to \infty} [(1 + r/m)^m - 1)] = \lim_{m \to \infty} \left[\left(1 + 1/(m/r)^{m/r}\right)^r - 1\right] = e^r - 1. \qquad (6.4)$$

Thus, for n periods [3] and $r = 10\%$ interest

$$F = P\left(i_{eff} + 1\right)^n = P\left(e^r - 1 + 1\right)^n = P\left(e^{rn}\right) \qquad (6.5)$$
$$F = 500\left(e^{0.1 \times 3}\right)$$
$$F = \$674.93.$$

Equation (6.5) is also the form used for calculating the future worth for continuous compounding. Note that over the 3-year period, continuous compounding results in $24.93 (or nearly 4%) more than simple compounding and earns $9.93 more than discrete compounding. The higher the interest rate, the greater the differences between the amounts of interest calculated by the methods as the calculations are not linear. Also, the more time periods, the greater the differences between the calculated values.

6.3   ACTUAL, COMPOUND, NOMINAL, AND EFFECTIVE ANNUAL INTEREST RATES

These four different interest rates and the differences between them will be discussed and illustrations of the differences will be presented.

The actual interest rate (i) represents the interest rate per compounding period. It is the most common of the interest rates used in engineering calculations. The actual interest rates can be expressed in different periods, such as:

12% per year	—interest would be compounded once per year
6% semi-annually	—interest would be compounded 2 times per year
3% per quarter	—interest would be compounded 4 times per year
1% per month	—interest would be compounded 12 times per year
0.03288% per day	—interest would be compounded 365 times per year

Nominal interest rate (r) represents the interest rate per year as obtained by multiplying the interest rate per period by the number of compounding periods per year. It is commonly known as the Annual Percentage Rate (APR) which is required for notifications on some loans. Since the other methods involve compounding more frequently, annual compounding results in the smallest value of the compound interest rates and simple interest is the lowest of all compounding methods.

Compound interest rate can be either discrete compounding or continuous compounding. Continuous compounding tends to give the largest amounts of interest. If the compounding period is one year, the continuous compounding rate is known as the annual effective interest rate. If the compounding rate is one year, the actual interest, nominal interest, and discrete compound interest will be the same as there is only one compounding period.

Let us consider a comparison of the interest rates for an interest rate of 3% per quarter for four periods. The values would be

Actual interest	$i = 3\%$/quarter for each quarter $= 0.03$/quarter
Nominal interest	$r = 4 \times 3\% = 12\% = 0.1200$/yr
Compound (Discrete)	$i_{eff\text{(discrete)}} = \left[(1 + .03)^4 - 1\right] \times 100 = 12.55\% = 0.1255$/yr
Compound (Continuous)	$i_{eff\text{(continuous)}} = [e^r - 1] \times 100$
	$\qquad = \left[e^{0.12} - 1\right] \times 100 = 12.75\% = 0.1275$/yr.

The effective interest rate depends upon which of the compounding methods is used and is based upon a one-year period. Now we shall calculate the amounts obtained after 3% per quarter on an initial amount of $1,000 compounded for three years. This results in 12 compounding periods. The calculations that are used are simple interest, compound (discrete) interest, and compound (continuous) interest. The actual interest rate is used in the calculation of the compound (discrete) multi-period calculations and the nominal period interest used in the calculation of the compound (continuous) multi-period calculations.

From Equation (6.2) for simple interest:

$$F = P(1 + i \times n) = 1,000(1 + 12 \times 0.03) = \$1,360.00 \text{ (Interest total is \$360.00)}.$$

Using Equation (6.3) for compound (discrete) interest calculations:

$$F = P(1 + i)^n = 1,000(1 + 0.03)^{12} = \$1,425.76.$$

(Interest total is $425.76 and more than 18% greater than simple interest.)

Using Equation (6.5) for compound (continuous) interest calculations:

$$F = P(e^{rn}) = 1,000 \left(e^{0.03 \times 12}\right) = \$1,433.33.$$

(Interest total is $433.33 which is more than 20% greater than simple interest.)

It is strongly advised to use the actual compounding period (a quarter or 3 months) and the corresponding interest (3%) as the annual interest rate is valid only for annual compounding periods. When other continuous expressions are used which have other factors such as $(e^r - 1)$ the results would not be correct.

The difference in the values of the two compound interest methods, discrete and continuous compounding, is small compared to their differences with the simple interest method. The more frequently the compounding, the smaller the difference between the two compounding (discrete and continuous) methods, but the differences increase as the interest rate increases and/or as the total investment time increases.

6.4 FACTORS IN DETERMINING INTEREST RATES

The interest rate considered as a basis for engineering and project calculations is the market interest rate. One interest rate is the prime interest rate, which is available to banks has been as low as 0.25% and is usually between 1% and 2% during normal times. The current rate by banks for deposits is between 0.5–2.0% but has been 3–5% in better economic times. The interest rate for automobile purchases is in the 3–6% range and had been as high as 8–10% range previously. When companies are considering returns on their investments, they typically want 10–20% return on average as some projects may make 30–40% return whereas other projects will lose money. Some of the major factors in considering an interest rate or rate of return are:

1. Administrative Expenses (1–5%);

2. Pure Gain or Profit (3–20%);

3. Risk of Inflation (1–200%); and

4. Risk of Loss (1–10%).

The risk of inflation is greater for investments when financial strife occurs in countries and prices rise rapidly. Construction projects can have rising labor and material prices and the total costs can go up rapidly. A major problem of inflation is that it is hard to predict long term and this has resulted in the use of the inflation-free interest rate.

6.5   INFLATION-FREE INTEREST RATES, CONSTANT CURRENCY, AND ACTUAL CURRENCY

The interest rates that have been considered have inflation as a component in the interest rate. The currency considered is the amount associated with a cash flow at the point of time at which it occurs and this is referred to as actual currency or current currency. The term currency has been used as dollars in the previous examples, but it could be the currency of any country.

Constant currency, or dollars, are dollars expressed in terms of the same purchasing power relative to a specific point in time, usually a base year. They represent the hypothetical purchasing power of future receipts and disbursements in terms of the purchasing power at the base year. Constant dollars are referred to as inflation free dollars and are often used on construction projects and government projects where the projects have a long life and estimating the inflation rates over a long period of time is highly speculative. The relationship between constant and actual currency is:

$$\text{Constant Currency (\$)} = [\text{Actual Currency at time } n \text{ (\$)}]/(1 + f)^n, \tag{6.6}$$

where

f = inflation rate (as a decimal) at time period n years in the future.

Constant currency is referenced to a base year, which is normally considered time zero or the beginning of the investment. Other names are constant dollars, real currency, inflation-free currency, and today's currency. Constant currency is typically used in construction or government projects having a project life 10 years or more and involving life cycle costs where the maintenance, repair, and rehabilitation values are difficult to predict that far in the future. For example, the life of a highway bridge can be 100 years and to predict the costs of a bridge deck replacement 20, 40, 60, and 80 years in the future is extremely difficult with any degree of accuracy, but the cost of a deck replacement today could be predicted with great accuracy. Thus, using constant currency, the replacement costs would be considered as the same as that of today. The interest rate used for discounting would not include the effects of inflation.

Although Equation (6.6) assumes the inflation will be constant over the n years of the investment, it typically will be changing and either an average value will be assumed or the inflation must be adjusted each year and would make the calculations more complex.

Actual currency or actual dollars are used in most applications, especially when the project investment life is under 20 years. Other names of actual dollars are nominal currency or current currency. The interest rate used is the effective or market interest which includes the effects of inflation. For most concerns in manufacturing, commercial, and project with short durations, actual currency or actual dollars are used.

The inflation-free interest rate, i^{if}, can be determined from the market interest rate and the inflation rate by:

$$(1 + i) = (1 + f)(1 + i^{if})$$

or directly by

$$i^{if} = [(1 + i)/(1 + f)] - 1, \qquad (6.7)$$

where

i = market interest rate (decimal)
i^{if} = inflation free interest rate (decimal)
f = inflation rate (decimal).

For example, if the market interest rate is 7.1% and the inflation rate was 2%, what would the inflation free interest rate be? Using Equation (6.7), one obtains:

$$i^{if} = [(1 + 0.071)/(1 + 0.02) - 1] = [1.05 - 1] = 0.05 \text{ or } 5\%.$$

6.6   CURRENCY EXCHANGE CALCULATIONS

The world is an international market and global projects must consider exchange rates involving different currencies. A project may be estimated in one currency, but performed in another country with a different currency. The fluctuations in currency rates can be large due to different inflation rates in the two countries. The exchange rate is the amount of one countries currency that would purchase one unit of another country's currency.

For example, if in the year 2015 when one U.S. dollar would purchase approximately 0.80 Euros, a U.S. investor invested $1,000 in Euros. In 2020, the investor decided to convert the Euros back to U.S. dollars and the exchange rate is one U.S. dollar to purchase 0.70 Euros. Consider Currency 1 as the original currency and currency rate in terms of amount of Currency 2 per unit of Currency 1. What did the investor receive?

Currency 1 (current value) = Currency 1 (original value)
$$\times \text{ [Currency Rate 2 (original)/Currency Rate2 (now)]}. \quad (6.8)$$

Thus,

Currency (now) = $1,000 × [(0.8 Euro/$1)/(0.7 Euro/$1] = $1,000[0.8/0.7] = $1,142.80.

If in the year 2015 one U.S. dollar purchased approximately 15 Pesos, a U.S. investor purchased $1,500 worth of Pesos. In 2020, the investor decided to convert the Pesos back to U.S. dollars and the exchange rate was 1 U.S. dollar purchased 25 Pesos. What did the investor receive?

Currency (now) = $1,500 × [(15 pesos/$1)/(25 pesos/$1)] = $1,500[15/25] = $900.

Thus, the effects of currency exchange rates can be rather large and must be considered in international projects. To reduce the problems, one usually does the cost and budget analysis in currency with the lowest inflation rate of the country where the project occurs or in the country where the project is managed and funded.

6.7   SUMMARY

Several types of interest rates have been presented. The interest rate most commonly applied is the market interest rate, but for long-term projects the inflation-free interest rate is used. The compound discrete interest rate is most frequently applied in calculations and is the effective interest most considered. However, the actual interest and the nominal interest rates are the basis for determining the effective interest rates for discrete compounding and continuous compounding. The actual dollars and market interest rate are used for most short-term projects whereas the constant currency and inflation-free interest rate are used for long-term projects. Currency evaluations are important when the project is being funded in one country and being constructed in another country as the inflation rates can be quite different.

6.8   REFERENCES

[1] Cresse, Robert C. and Kumar, Pradeep, *Engineering Economy Basics*, 1.5 CEU, p 147, May 18–19, 2000. 77

[2] Cresse, Robert C. and Kumar, Pradeep, *Intermediate Engineering Economics*, 1.5 CEU, p 147, June 28–29, 2001.

[3] Creese, Robert C., Engineering economics for engineers, estimators, managers and project anagers, *AACE International Annual Meeting*, 1.6 CEU, p. 94, June 28–29, 2008. 77, 80

[4] Creese, Robert C. and Adithan, M., *Strategic Cost Analysis for Project Managers and Engineers*, New Academic Science Limited, Tunbridge Wells, UK, p. 187, 2012. 77, 79

6.9   EVALUATIVE QUESTIONS

1. What is the value of $100,000 invested at 10% simple interest per year for 10 years?

2. What is the value of $100,000 invested at 10% annual discrete compound interest rate per year for 10 years?

3. What is the value of $100,000 invested at 10% annual continuous compound interest per year for 10 years?

4. The interest rate is 2% per month and the time period is 1 year.

    (a) What is the nominal interest rate?

    (b) What is the discrete effective interest rate?

    (c) What is the continuous effective interest rate?

5. A person invests $50,000 at 9% simple interest for 4 years. At the end of the 4 years, the entire amount (principal and interest) is invested at 11% discrete compounded annually for 10 years. What is the amount of the investment after 14 years?

6. The inflation rate is 4% and the inflation free interest rate is 8%, what is the market interest rate?

7. The market interest rate is 20% and the inflation rate is 5%, what is the inflation free interest rate?

8. If continuous compound interest is used and the nominal rate is 15%, what is the equivalent discrete annual effective interest rate?

9. If the discrete annual effective interest rate is 20%, what is the equivalent nominal rate for the continuous compound interest.

10. If the currency in the country where the project is being constructed (Country C) has an inflation rate of 10% and the country where the project is being managed and funded (Country F) has an inflation rate 15%, what currency should the funds of the project be kept and why?

11. The $10,000,000 project in Country B is being funded by country A. The exchange rate of Country A is 2 currency units per Country B currency unit. It is projected that the exchange rate for the last half of the project ($5,000,000) will drop and the exchange rate of Country A will drop to 1.5 currency units per Country B currency unit. Which country should the project funded currency be located and why?

12. Canada had an exchange rate of 0.95 Canadian dollars per U.S. dollar in 2015 and a U.S. investor spent $10,000 U.S. dollars to purchase Canadian dollars. How many Canadian dollars did he receive? In 2020 he exchanged those Canadian Dollars for U.S. dollars at the exchange rate of 0.70 Canadian dollars per U.S. dollar. How many U.S. dollars did he receive in 2020?

CHAPTER 7

Basic Mathematical Relationships for Economic Calculations

7.1 INTRODUCTION

The following chapters will derive and illustrate the fundamental relationships used for economic evaluations. The derivation of the basic economic formulas are based upon a few algebraic relationships and applying these relationships will make the understanding of the economic formulas easier. The mathematics involved in the derivations of the formulas in economics are not complex and some basic references are listed [1, 2] The following sections will focus on the expressions for the sums of numbers, the arithmetical progression, the geometric progression, and the infinite limit expression.

7.2 SUMS OF NUMBERS

In evaluating projects one typically examines all the revenues in the future time periods and evaluates them at the beginning of the project to determine the expected cost. Thus, if one wants to determine the total amount or sum S of an amount z each period for n periods, the total would be:

$$S(n) = \sum_{i=1}^{n} z_i = nz. \tag{7.1}$$

Let us suppose that $n = 5$ and $z = 6$, then the sum is

$$S(5) = 6 + 6 + 6 + 6 + 6$$
$$= 5 \times 6 = 30.$$

Now let us assume that instead of the period amount remaining constant that it increases by the same amount each period (as in a uniform gradient); that is for the second period it is $2z$, for the third period it is $3z$, and for the nth period it is nz. This is the basic sum of numbers expression. The total amount S for the n periods would be:

$$S(n) = \sum_{i=1}^{n} nz = zn(n + 1)/2. \tag{7.2}$$

Let us suppose the $n = 5$ and $z = 6$ which implies the amount for the second period, $S(2) = 2 \times 6 = 12$, and the total amount S for the 5 periods would be:

$$S(5) = 6 + 12 + 18 + 24 + 30 = 90$$
$$= 6 \times 5 \times (5 + 1)/2 = 90.$$

The arithmetical progression is a modification of the basic sum of numbers where the increment y is different than the base amount z. The increment y starts in the second period. The total amount S for the n periods would be:

$$S(n) = \sum_{i=1}^{n} z + \sum_{i=2}^{n} y = nz + n(n-1)y/2. \tag{7.3}$$

Let us suppose that $n = 5$, $z = 6$, and $y = 3$ and the total amount S for the n periods would be:

$$S(5) = 6 + 9 + 12 + 15 + 18 = 60$$
$$= 5 \times 6 + 5 \times 4 \times 3/2 = 30 + 30 = 60.$$

7.3   GEOMETRIC PROGRESSION

The expression used in the derivation of discrete interest economic expressions is the geometric progression, commonly called the geometric series, when dealing with discrete interest relationships. The discrete relationships are those where there is a discrete payment at a discrete interest rate at a specific time. This discrete interest rate implies that it is compounded at a fixed time period, whereas a continuous interest implies that it is compounded continuously over time. The case of discrete payments and discrete interest rates is the most common of the economic problems.

The basic mathematical relationship for discrete interest problems is the geometric series and for continuous interest problems the infinite limits expression is used. The geometric series expression is:

$$S = a + aR + aR^2 + aR^3 + aR^4 + \cdots + aR^{(n-1)} = a\left[(R^n - 1)/(R - 1)\right], \tag{7.4}$$

where

    S = sum of the series of n terms
    a = constant which occurs in all terms
    R = ratio between terms
    n = number of terms in the sum (including the initial term without the ratio).

Consider the following example where one takes the number $4(a = 4)$ and doubles it $(R = 2)$ for 3 additional periods $(n = 3 + 1 = 4)$ and determine the sum. That is:

$$S(5) = 4 + (4 \times 2) + \left(4 \times 2^2\right) + \left(4 \times 2^3\right) = 4 + 8 + 16 + 32 = 60.$$

If one uses the geometric series expression, there are four terms (the initial term and the 3 additional periods), so

$$n = 4$$
$$R = 2$$
$$a = 4 \quad \text{and thus}$$
$$S(4) = 4 \times \left[2^4 - 1\right]/[2 - 1] = 4 \times [16 - 1]/[2 - 1] = 4 \times [15]/[1] = 60.$$

This becomes very useful when n is large. In most interest calculations, the ratio is $R = (1 + i)$ which is the periodic compounding amount for compound discrete interest calculations. This will be used in the derivation of the discrete compounding factors in the various economic expressions.

Let us consider the investment of \$2,000 per year at an annual interest rate of 1% for 5 years. Thus,

$$a = \$2,000$$
$$R = 1 + i = 1.01$$
$$n = 5$$
$$S(n) = a\left[(R^n - 1)/(R - 1)\right]$$
$$S(5) = \$2,000\left[(1.01)^5 - 1\right]/(1.01 - 1) = 2,000[1.05101 - 1]/[1.01 - 1]$$
$$= 2,000 \times 5.101005$$
$$= \$10,202.01.$$

Thus, the total interest over the 5-year period is \$202.01 and the effect of compounding is small, only \$2.01 If the interest rate was 3% ($R = 1.03$),

$$S(n) = a\left[(R^n - 1)/(R - 1)\right]$$
$$S(5) = \$2,000\left[(1.03)^5 - 1\right]/(1.03 - 1) = 2,000 \times 5.309136 = \$10,618.27.$$

The total interest over the 5-year period is \$618.27 and the effect of compounding is \$18.27, which is much larger than the \$2.01 or even more than 3 times the \$2.01 which is \$6.03. The higher the interest rate, the greater the effect of compounding and the greater the time period also increases the compounding effect.

7.4   INFINITE LIMIT

The infinite limit is used for continuous interest problems, that is when the interest is compounded continuously over the periods rather at the end of the discrete time periods. The continuous compounding is also used when continuous cash flows are considered instead of discrete cash flows. However, continuous cash flow analysis is rarely used and will not be presented in detail. The continuous interest expression is based on:

$$Lim_{k \to \infty} [1 + 1/k]^k = e = 2.718281834.$$

This can be illustrated by taking values of k such as 1, 2, 3, 100, 1,000, and 10,000 and note that the value approaches the limit of e, which is 2.718 to 3 decimals:

$$Limit = [1 + 1/1], \left[1 + \frac{1}{2}\right]^2, [1 + 1/3]^3, \ldots [1 + 1/100]^{100}, \ldots [1 + 1/1000]^{1000}, \ldots$$

$$\ldots [1 + 1/10000]^{10000}$$

$$Limit = \quad 2, \qquad 2.25, \qquad 2.37, \qquad 2.704, \qquad 2.71692,$$

$$2.71815.$$

The infinite limit was used in the derivation of continuous compound interest by letting the number of discrete compounding periods go to infinity; that was:

$$i_{eff} = \lim_{m \to \infty} [(1 + r/m)^m - 1] = \lim_{m \to \infty} \left[\{1 + 1/(m/r)\}^{m/r}\right]^r - 1 = e^r - 1$$

$$i_{eff} = e^r - 1, \tag{7.5}$$

where

i_{eff} = interest rate (effective interest rate on an annual basis)
r = nominal interest rate.

One can also express the relationship to determine the nominal interest rate equivalent to the annual effective interest rate as:

$$e^r = \left(i_{eff} + 1\right) \tag{7.6}$$

or

$$r = \ln\left(i_{eff} + 1\right). \tag{7.7}$$

Equation (7.5) is frequently used to convert discrete compounding factors into continuous compounding factors and Equations (7.6) and (7.7) to determine annual nominal interest r from the continuously compounded interest rate.

7.5   SUMMARY

The basic mathematical expressions for sums of numbers, including the basic sum of numbers expression, the arithmetic progression, the geometric progression, and the infinite limit. The geometric series expression will be used in the development of the discrete interest compounding expressions and the infinite limit will be used to convert the discrete interest formulas to continuous interest formulas. These expressions are relatively simple and are the basis of the economic expressions presented in the following chapters.

7.6   REFERENCES

[1] Hodgman, Charles D., Ed., *Mathematical Tables from Handbook of Chemistry and Physics*, 10th ed., Chemical Rubber Publishing Co., Cleveland, OH, pp. 294–296, 1954. 87

[2] Creese, Robert C. and Adithan, M., *Strategic Cost Analysis for Project Managers and Engineers*, New Academic Science Limited, Tunbridge Wells, UK, pp. 36–38, 2012. 87

7.7   EVALUATIVE QUESTIONS

1. What is the sum of the numbers from 1 to 10?

2. What is the sum of the numbers from 40 to 50?

3. A gradient of 3 is to be summed over 10 periods and determine the total sum.

4. A cost of $40 per period is for wages. The cost for materials is $5 per period and increases at the rate of $1 per period. What is the total wage cost for 10 periods? What is the total material cost for 10 periods?

5. John earns $5 and the amount is doubled each period for the next 6 periods, what is the sum John will have in his account.

6. Mary earns $10 and the amount is tripled each period for the next 4 periods, what is the sum Mary will have at the end of these 5 periods?

7. Francisco saves $100 per month and his interest rate is 1/2% compounded monthly. What amount will Francisco have at the end of the year?

8. Juanita has invested 10,000 Pesos in a high risk bond which pays 1% compounded monthly which is reinvested automatically. What is the expected value of the bond and total interest at the end of the year?

9. An investment of $2,000 is made and the interest rate is 0.5% per month. At the end of a year, determine the total amount available if:

    (a) simple interest is used.

    (b) discrete compound interest is used.

    (c) continuous compound interest is used.

Also determine

   (d) the nominal interest rate.

   (e) the effective discrete compound interest rate.

   (f) the effective continuous compound interest rate.

10. An investment of $2,000 is made and the interest rate is 1.5% per month. At the end of a year, determine the total amount available if:

   (a) simple interest is used.

   (b) discrete compound interest is used.

   (c) continuous compound interest is used.

Also determine

   (d) the nominal interest rate.

   (e) the effective discrete compound interest rate.

   (f) the effective continuous compound interest rate.

11. An investment of $2,000 is made and the interest rate is 1.5% per quarter. At the end of a year, determine the total amount available if:

   (a) simple interest is used.

   (b) discrete compound interest is used.

   (c) continuous compound interest is used.

Also determine

   (d) the nominal interest rate.

   (e) the effective discrete compound interest rate.

   (f) the effective continuous compound interest rate.

12. An investment of $2,000 is made and the interest rate is 1.5% per quarter. At the end of five years, determine the total amount available if:

   (a) simple interest is used.

   (b) discrete compound interest is used.

   (c) continuous compound interest is used.

   (d) the nominal interest rate.

   (e) the effective discrete compound interest rate.

   (f) the effective continuous compound interest rate.

CHAPTER 8

Basic Economic Factors and Equations

8.1 INTRODUCTION

There are several expressions utilized in economics to determine the worth of various payments types over time for the evaluation of projects. These are evaluated as Present Worth values, which is the value of the payments at the start or time zero, or Future Worth values, which is the value of the payments at a specific future time, typically at the end of the project life. The payments can be a single payment or a uniform series of payments. The expressions are generally divided into two categories: the basic expressions and the gradient expressions. The basic expressions will be examined in this chapter and the more complex gradient expressions will be examined in the next chapter. The primary reference for this section is from a previous short course and a previous book [1], but the materials can be found in numerous standard engineering economics textbooks [2, 3] giving slightly different approaches and additional problems. Some of the expressions developed is this and the following chapters are available using Excel [4, 5], but one should first use and program the expressions before relying on packaged software expressions.

The two categories of the basic economic expressions are classified by the type of payments—the single payment expressions and the uniform payment expressions. There are two single payment expressions and four uniform payment expressions and all six will be presented in this chapter. The expressions will be first developed for the discrete compound interest rate and discrete payment case. The developed expressions will then be modified for the continuous compound interest rate with discrete payment case. The highly advanced case of continuous compound interest rate and continuous payments will not be presented as it is currently rarely applied.

8.2 SINGLE PAYMENT DISCRETE INTEREST FACTORS

The two single payment cases developed with discrete interest and discrete payments are the future worth expression and the present worth expression. The future worth expression will be presented first as it is easier to illustrate.

8.2.1 DISCRETE INTEREST FUTURE WORTH FACTOR $(F/P, i, n)$

The future worth expression converts a present amount to a future worth for a given interest rate and given number of compounding periods.

The notation used will be:

F = future worth (total amount after n compounding periods)
P = present worth (total amount at time zero, the beginning of the study period)
i = interest rate per compounding period (discrete compound interest)
I = interest earned in compounding period = $i \times P$
n = number of compounding periods
$\uparrow$ = cash flow is occurring.

A graph indicating where the present worth and future worth are with respect to time appears in Figure 8.1.

Figure 8.1: Present and future worth with respect to time.

Table 8.1 illustrates the calculations starting with the investment P (Present Worth) at the start of the time period and Future Worth F at the end of the time period. This figure is similar to that in Chapter 6 to obtain Equation (6.3). In most instances the future worth is greater than the present worth.

Thus, from Table 8.1 it can be seen that the value at the end, the Future Worth Equation is:

$$F = P \times (1 + i)^n \tag{8.1}$$

where

F = future worth (total amount after n compounding periods)
P = present worth (total amount at time zero, the beginning of the study period)
i = interest rate per compounding period (discrete compound interest)
n = number of compounding periods.

The conversion factor to convert the present worth to the future worth is called the single payment future worth discrete compound amount factor is designated as $(F/P, i, n)$, is stated as F given P, i, n, and is equal to $(1 + i)^n$. The discrete future worth factor is:

$$(F/P, i, n) = (1 + i)^n. \tag{8.2}$$

Table 8.1: Payment illustration for future worth expression derivation

Time (End of Period)	Present Worth (P) at Start of Period		Interest (I) Earned During Period		Future Worth is the Principal + Interest Total at End of Period
1	P	+	IP	=	$P \times (1 + i)$
2	$P \times (1 + i)$	+	$IP \times (1 + i)$	=	$P \times (1 + i)^2$
3	$P \times (1 + i)^2$	+	$IP \times (1 + i)^2$	=	$P \times (1 + i)^3$
4	$P \times (1 + i)^3$	+	$IP \times (1 + i)^3$	=	$P \times (1 + i)^4$
...				=	
n	$P \times (1 + i)^{n-1}$	+	$IP \times (1 + i)^{n-1}$	=	$P \times (1 + i)^n = F$

8.2.2 DISCRETE INTEREST FUTURE WORTH EXAMPLE

If a single payment of $100,000 is invested at 15% interest compounded annually, the compound amount at the end of the 4th year would be:

$$F = \$10,000(1 + 0.15)^4 = \$10,000 \times 1.74901 = \$17,490$$

and

$$(F/P, i = 15\%, n = 4) = (1 + i)^n = (1 + 0.15)^4 = 1.74901$$

so

$$F = P \times (F/P, i = 15, n = 4) = \$10,000 \times 1.74901 = \$17,490.$$

The $(F/P, i, n)$ factor is given for the formulas in Table 8.2 and the Appendix and one should use the formulas to calculate the values. Students should learn to calculate and program the formulas rather than use tables as many interest rates are not available in the tables.

The total interest earned, I (Total), would be the difference between F and P and would be:

$$I(\text{Total}) = F - P = \$17,490 - \$10,000 = \$7,490.$$

8.2.3 DISCRETE INTEREST PRESENT WORTH FACTOR $(P/F, i, n)$

The present worth expression is the inverse of the future worth expression; that is, the expression is:

$$P = F(1 + i)^{-n} = F/(1 + i)^n, \tag{8.3}$$

where

P = present worth (initial amount at time zero)

F = future worth (total amount after n time periods)
i = discrete interest rate per compounding period
n = number of compounding periods.

The conversion factor to convert the future worth to the present worth is called the single payment present worth compound amount factor, is designated as $(P/F, i, n)$, is stated as P given F, i, n, and is equal to $(1 + i)^{-n}$ or $1/(1 + i)^n$. That is:

$$(P/F, i, n) = (1 + i)^{-n}. \tag{8.4}$$

8.2.4 DISCRETE PRESENT WORTH EXAMPLE

If $10,000 is desired at the end of 4 years and the discrete interest rate is 15% compounded annually, what amount would one need to be deposited initially?

$$P = F/(1 + i)^n,$$

therefore

$$P = \$10,000/(1 + 0.15)^4 = \$10,000/1.74901 = \$5,718$$

and

$$(P/F, i, n) = 1(1 + i)^{-n} = 1/(1 + 0.15)^4 = 0.57175$$

so

$$P = \$10,000 \times (P/F, i = 15, n = 4) = \$10,000 \times 0.57175 = \$5,717.5 = \$5,178.$$

The $(P/F, i, n)$ factor is listed in Table 8.2 at the end of this chapter and in the Appendix. The present worth analysis is often referred to as "discounting," that is bringing future values back to the present value. As the sample indicates, the future worth of $10,000 at the end of four years in the future is worth only $5,178 when discounted by 15% per year to the current year.

8.3   UNIFORM SERIES PAYMENTS DISCRETE INTEREST FACTORS

The four uniform payment factors with discrete interest payments to be developed are the uniform series future worth factor $(F/A, i, n)$, the sinking fund factor $(A/F, i, n)$, the uniform series present worth factor $(P/A, i, n)$, and the capital recovery factor $(A/P, i, n)$. All of these factors involve the payment of an amount A at the end of every period. In the development of the various economic expressions, the assumption is made that the discrete payment is at the end of the payment period, not at the beginning of the payment period. A simple modification can be made to convert end of period payments to beginning of the period payments.

8.3.1   UNIFORM SERIES DISCRETE INTEREST FUTURE WORTH FACTOR $(F/A, i, n)$

(Also called **Discrete Compound Amount Factor.**)

Payments of the amount A are made at the end of the period for n periods (but that does not include time 0) to determine a total amount F at the end of the last period. The discrete interest rate i will be compounded at the end of each period and the payment sum plus the interest accumulated will result in a future worth of F at the end of the n periods. This includes a period payment A in the final period to complete the future worth value of F as the system is based upon end-of-period payments. The geometric series expression is used to develop the formulas. A graphical representation of the payments and future worth appears in Figure 8.2.

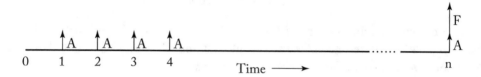

Figure 8.2: A graphical representation of the payments and future worth. Where $A =$ uniform payment each period n, $F =$ future worth (amount at time n), $i =$ interest rate per compounding period, and $n =$ number of compounding periods.

The expression can be derived using the geometric series expression by the payments starting from the last period back to the first period:

$$F = A \text{ (end of last period)} + A(1+i) + A(1+i)^2 + \cdots + A(1+i)^{(n-1)} \text{ (end of first period)}$$
$$= A \left(1 + (1+i) + (1+i)^2 + \ldots + (1+i)^{(n-1)}\right)$$
$$= A \text{ (geometric series)}.$$

If one lets $r = (1+i)$, one has

$$F = A \, (r^n - 1) / (r - 1) = A \, [(1+i)^n - 1] / [(1+i) - 1] = A \, [(1+i)^n - 1] / [i]$$
$$F = A \, [(1+i)^n - 1] / [i] \tag{8.5}$$

and

$$F = A \, (F/A, i, n) = A \, [(1+i)^n - 1] / [i]$$

thus

$$(F/A, i, n) = [(1+i)^n - 1] / [i]. \tag{8.6}$$

8.3.2 UNIFORM SERIES DISCRETE INTEREST FUTURE WORTH EXAMPLE

What is the future worth of $1,000 deposited at the end of the year each year for 10 years when the interest is 10%?

$$F = \$1{,}000 \times \left[(1 + 0.10)^{10} - 1\right] / [0.10] = \$1{,}000 \times [1.5937/0.10] = \$1{,}000 \times [15.937]$$
$$= \$15{,}937.$$

The total amount of interest earned over the 10 years would be:

$$I\ (\text{total}) = F - n \times A$$
$$= 15{,}937 - 10 \times 1{,}000 = \$5{,}937.$$

The compound amount factor $[F/A, i = 10\%, n = 10] = 15.937$ may be listed in the tables, but if one had an interest rate of 9.8% it would not be available and thus one should be able to use the compound amount factor expression and thus:

$$[F/A, i = 9.8\%, n = 10] = \left[(1 + 0.098)^{10} - 1\right] / [0.098] = 15.785.$$

Thus,

$$F = A \times (F/A, i, n) = \$1{,}000 \times [F/A, i = 9.8\%, n = 10] = \$1{,}000 \times 15.785 = \$15{,}785.$$

8.3.3 SINKING FUND DISCRETE INTEREST FACTOR $(A/F, i, n)$

The sinking fund determines the amount of A to obtain a desired amount F in the future. The sinking fund factor involves the same terms as the compound amount factor and the graphical representation of the payments and future worth is the same and is repeated as in Figure 8.3.

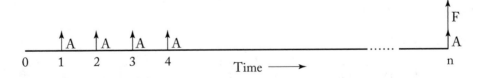

Figure 8.3: A graphical representation of the payments and future worth.

The Sinking Fund Factor is the inverse of the Compound Amount Factor and thus it would be:

$$(A/F, i, n) = 1/(F/A, i, n).$$

Therefore, the Sinking Fund Equation is:

$$A = F \times \{i / [(1 + i)^n - 1]\} \tag{8.7}$$

and

$$(A/F, i, n) = 1/\left[(1+i)^n - 1\right]/[i]$$
$$= i/\left[(1+i)^n - 1\right],$$

(8.8)

where

A = uniform payment in each period n
F = future worth (amount at time n)
i = interest rate
n = number of compounding periods.

8.3.4 SINKING FUND DISCRETE INTEREST FACTOR EXAMPLE

Melania wants to have $1,000 at the end of 10 years, so what amount would she have to save at the end of each year if the interest rate over the 10-year period is 15% to obtain the desired $1,000?

$$A = \$1,000\left[0.15/(1+0.15)^{10} - 1\right] = \$1,000 \times [0.15/(3.046)]$$
$$= \$1,000 \times (0.04925) = \$49.25.$$

Thus, $(A/F, i = 15\%, n = 10) = 0.04925$.

Thus, Melania must deposit $49.25 at the end of each year for 10 years to have $1,000 at the end of the ten year period. Her final payment at period n is necessary to make the total one thousand dollars.

8.3.5 UNIFORM SERIES DISCRETE INTEREST PRESENT WORTH FACTOR $(P/A, i, n)$

The uniform series present worth factor is an extension of the present worth expression to have payments in each period and not only in the last period. The uniform series present worth factor is used to convert a uniform series of n payments of the amount A at an interest rate i to a present worth amount P. A graphical represent of the payments and the present worth is presented as in Figure 8.4.

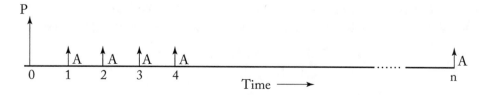

Figure 8.4: Payments and the present worth.

To determine the present value P, each payment A must be discounted at the interest rate i by the number of periods it occurs in the future. The discount factor is $(1 + i)^{-n}$ for each period where n is the specific period for that particular payment. The expression can be derived using the geometric series expression by:

$P = A$ discounted 1 period $+ A$ discounted 2 periods $+ \cdots + A$ discounted n periods

$P = A/(1 + i) + A/(1 + i)^2 + A/(1 + i)^3 + A/(1 + i)^4 + \cdots + A/(1 + i)^n$

$P = A/(1 + i)$

$$\left[1 + 1/(1 + i) + 1/(1 + i)^2 + 1/(1 + i)^3 + 1/(1 + i)^4 + \cdots + 1/(1 + i)^{(n-1)}\right]$$

[geometric series].

Now using the geometric series equation of Equation (7.4) where $a = A/(1 + i)$, $S = P$ and $R = (1 + i)^{-1}$ or $1/(1 + i)$ one obtains:

$$S = a\left[(R^n - 1)/(R - 1)\right]$$

$$P = A/(1 + i) \times \left[(1/(1 + i))^n - 1\right]/\left[(1/(1 + i) - 1)\right]$$

$$= A/(1 + i) \times \left[\{1 - (1 + i)^n\}/(1 + i)^n\right]/\left[\{1 - (1 + i)\}/(1 + i)\right]$$

$$= A \times \left[\{1 - (1 + i)^n\}/(1 + i)^n\right]/[-i]$$

$$P = A \times \left[\{(1 + i)^n - 1\}/(i(1 + i)^n)\right], \tag{8.9}$$

where

$A = $ uniform payment each period n

$P = $ present worth (amount at time zero)

$i = $ interest rate

$n = $ number of compounding periods.

The uniform series present worth factor is thus

$$(P/A, i, n) = \left[\{(1 + i)^n - 1\}/(i(1 + i)^n)\right]. \tag{8.10}$$

8.3.6 UNIFORM SERIES DISCRETE INTEREST PRESENT WORTH EXAMPLE

Barack won a lottery which promised \$20,000,000 paid as \$1,000,000 at the end of each year for 20 years. If the interest rate is 15%, what is the present worth of the Barack's lottery winnings?

$$P = A \times (P/A, i = 15\%, n = 20)$$

$$P = \$1,000,000 \times \left[(1.15)^{20} - 1\right]/\left[0.15 \times (1.15)^{20}\right]$$

$$= \$1,000,000 \times (6.2593)$$

$$= \$6,259,300.$$

The total present worth amount of the lottery is less than one-third of the listed $20,000,000 prize amount when the interest rate is 15%. From the calculations, one notes that:

$$(P/A, i = 15\%, n = 20) = 6.2593.$$

8.3.7 CAPITAL RECOVERY DISCRETE INTEREST FACTOR $(A/P, i, n)$

The Capital Recovery Discrete Interest Factor is the inverse of the Uniform Series Present Worth Discrete Interest Factor and has the same graphical representation as the Uniform Series Present Worth Factor. It is the amount that one must have initially in order to receive an amount A at the end of each year for n years when the interest rate is i during the n year period (Figure 8.5).

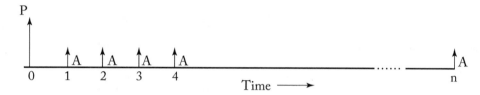

Figure 8.5: Uniform series present worth factor.

The solution is to determine the amount of the payment A based upon the values of P, i, and n. The Capital Recovery Factor is the inverse of the Uniform Series Present Worth Factor and is:

$$A = P \times [(i(1 + i)^n)] / [(1 + i)^n - 1]. \tag{8.11}$$

where

$A =$ uniform payment each period n (does not include time zero)
$P =$ present worth (amount at time zero)
$i =$ interest rate
$n =$ number of compounding periods.

The capital recovery factor is thus:

$$(A/P, i, n) = [(i(1 + i)^n)] / [(1 + i)^n - 1]. \tag{8.12}$$

8.3.8 CAPITAL RECOVERY DISCRETE INTEREST FACTOR EXAMPLE

Michelle has purchased a new automobile for $50,000 after a down payment of $5,000, that is, the total cost was $55,000. Her loan is for $50,000 for a period of 3 years at an interest rate of 15%/year. She wants to pay off the loan at the end of 3 years.

(a) If annual payments are made, what is the annual payment?

$$A = \$50,000 \times [0.15(1.15)^3] / [1.15^3 - 1] = \$50,000 \times [0.437980] = \$21,899$$

(b) The total interest paid over the three year period is:

$$I \text{ (total)} = n \times A - P = 3 \times \$21{,}899 - \$50{,}000 = \$15{,}697.$$

(c) If monthly payments are made, what is the monthly payment?

The interest rate would be 1.25% (15%/12) per month and $n = 36$

$$A = \$50{,}000 \times \left[0.0125(1.0125)^{36}\right] / \left[(1.0125)^{36} - 1\right] = \$1{,}754.93.$$

(d) The total interest paid over the 36 monthly payments would be

$$I \text{ (total)} = n \times A - P = 36 \times \$1{,}754{,}93 - \$50{,}000$$
$$= \$63{,}177.56 - \$50{,}000 = \$13{,}178.$$

Michelle could save \$2,519 in interest by paying monthly instead of yearly.

8.4   SINGLE PAYMENT CONTINUOUS INTEREST FACTORS

As indicated in Chapter 7, the effective interest expressions can be expressed in terms of the nominal interest r. Let the effective interest be expressed as i, then the expressions become:

$$i = e^r - 1 \tag{8.13}$$

and

$$(1 + i) = e^r \tag{8.14}$$

and similarly

$$(1 + i)^n = e^{rn}. \tag{8.15}$$

8.4.1   CONTINUOUS INTEREST FUTURE WORTH SINGLE PAYMENT FACTOR $(F/P, r, n)$

The discrete future worth equation and discrete future worth factor equations from Section 8.2.2 were:

$$F = P(1 + i)^n \tag{8.3}$$

and

$$(F/P, i, n) = ((1 + i)^n. \tag{8.4}$$

The continuous future worth factor and continuous future worth equations using Equation (8.15) and the discrete future worth factor and discrete future worth equations would be:

$$F = Pe^{rn} \tag{8.16}$$

and

$$(F/P, r, n) = e^{rn}, \tag{8.17}$$

where

$F =$ future worth (total amount after n time periods)
$P =$ present worth (initial amount at time zero)
$r =$ continuous interest rate per compounding period
$n =$ number of compounding periods.

8.4.2 CONTINUOUS INTEREST FUTURE WORTH SINGLE PAYMENT EXAMPLE

George has purchased a precious painting for $10,000 and expects it to appreciate in value at a nominal interest rate of 15% per year compounded continuously over the next 4 years. What would be the value of the painting at the end of 4 years that George would be expecting?

$$
\begin{aligned}
F &= Pe^{rn} \\
&= \$10{,}000 \times e^{\cdot 0.15 \times 4} \\
&= \$10{,}000 \times 1.82212 \\
F &= \$18{,}221
\end{aligned}
$$

and

$$(F/P, 15, 4) = e^{0.15 \times 4} = 1.8221.$$

If George is correct in his purchase assumptions, he would have a gain of $8,221 on his investment. Note that if discrete compounding was used, his gain would have been $7,490 and if simple interest was used his gain would have been only $6,000.

8.4.3 CONTINUOUS INTEREST PRESENT WORTH SINGLE PAYMENT FACTOR $(P/F, r, n)$

The discrete present worth factor and discrete present worth equations from Section 8.2.3 were:

$$P = F(1 + i)^{-n}$$

and

$$(P/F, i, n) = (1 + i)^{-n}.$$

The continuous present worth factor and continuous present worth equations using Equation (8.15) and the discrete present worth equation and discrete present worth factor equations would be:

$$P = Fe^{-rn} \tag{8.18}$$

and

$$(P/F, r, n) = e^{-rn}. \tag{8.19}$$

8.4.4 CONTINUOUS INTEREST PRESENT WORTH SINGLE PAYMENT EXAMPLE

Laura wants to have $10,000 for the purchase a new car in 4 years as her current car will be lucky to last that long. Her Aunt Barbara wants to buy a precious book from her. She wants to sell the book at a price that would appreciate to $10,000 at the end of 4 years and has an opportunity to invest in a bank note that would pay 15% interest with continuous compounding over the next 4 years. What is the price she needs to sell her precious book at to have the $10,000 in 4 years?

$$P = Fe^{-rn}$$
$$= \$10{,}000e^{-0.15 \times 4}$$
$$= \$10{,}000e^{-0.6}$$
$$= \$10{,}000 \times 0.54881$$
$$P = \$5{,}488.1$$

and

$$(P/F, r, n) = e^{-rn} = e^{-0.15 \times 4} = 0.54881.$$

If Laura can persuade Aunt Barbara to pay $5,488 for the book now and invest in the bank note for 4 years, she would have her desired $10,000. Note that if discrete interest was used instead of continuous, she would have needed Aunt Barbara pay $5,718 for the book.

8.5 UNIFORM SERIES PAYMENTS CONTINUOUS INTEREST FACTORS

The factors for the uniform series payments continuous interest are more complex than the single payment continuous interest factors, but they can be easily obtianed from the Uniform Series Payments Discrete Interest Factors and they will be illustrated by examples in the following sections.

8.5.1 UNIFORM SERIES CONTINUOUS INTEREST FACTORS–FUTURE WORTH, SINKING FUND, PRESENT WORTH, AND CAPITAL RECOVERY

The uniform series discrete future worth factor in Section 8.3.1 was:

$$(F/A, i, n) = \left[(1 + i)^n - 1\right] / [i].$$

Using Equations (8.13) and (8.15) and the nominal interest rate r replacing the discrete interest rate i, the uniform series continuous future worth factor becomes:

$$(F/A, r, n) = [e^{rn} - 1] / [e^r - 1]. \tag{8.20}$$

The uniform series continuous sinking fund continuous factor (uniform series continuous future worth factor) in Section 8.3.3 was:

$$(A/F, i, n) = [i] / [(1 + i)^n - 1].$$

Using Equations (8.13) and (8.15) and the nominal interest rate r replacing the discrete interest rate i, the uniform series continuous future worth factor becomes:

$$(A/F, r, n) = [e^r - 1] / [e^{rn} - 1]. \tag{8.21}$$

The uniform series discrete present worth factor in Section 8.3.5 was:

$$(P/A, i, n) = [(1 + i)^n - 1] / [i(1 + i)^n].$$

Using Equations (8.13) and (8.15) and the nominal interest rate r replacing the discrete interest rate i, the uniform series continuous present worth factor becomes:

$$(P/A, r, n) = [e^{rn} - 1] / [(e^r - 1) e^{rn}]. \tag{8.22}$$

Finally, the uniform series discrete capital recovery factor in Section 8.3.7 was:

$$(A/P, i, n) = [i(1 + i)^n] / [(1 + i)^n - 1].$$

Using Equations (8.8) and (8.11) and the nominal interest rate r replacing the discrete interest rate i, the uniform series continuous present worth factor becomes:

$$(A/P, r, n) = [(e^r - 1) e^{rn}] / [e^{rn} - 1]. \tag{8.23}$$

8.5.2 UNIFORM SERIES CONTINUOUS INTEREST FUTURE WORTH $(F/A, r, n)$ EXAMPLE

Uncle Bill wants to take a trip to Morgantown, WV from Hope, AK to watch a football game. The bus fare is $400 and he plans to put $100 per month into an account for 6 months (that is 1/2 year). He plans to use the additional monies for his room, meals, and the $85 dollar football ticket. If the account pays continuous annual interest of 6% (which is 1/2% per month), how much money will he have to spend for his room and meals. Using Equation (8.20), the total future amount would be:

$$(F/A, r, n) = [e^{rn} - 1] / [e^r - 1]. \tag{8.20}$$

Therefore,

$$F = A \, [e^{rn} - 1] / [e^r - 1] \tag{8.24}$$
$$F = 100 \, [e^{0.005 \times 6} - 1] / [e^{0.005} - 1]$$
$$= 100 \, [e^{0.03} - 1] / [e^{0.005} - 1]$$
$$= \$100[1.030454 - 1]/[1.0050125 - 1]$$
$$= \$100 \times 6.0756$$
$$= \$607.56.$$

Note that the low interest rate and short period gives a small amount of interest, only $7.56. Therefore, the money Uncle Bill would have for spending on his room and meals is:

$$\$607.56 - \$400 - \$85 = \$122.56.$$

If one had utilized $r = 6\%$/year and $n = 1/2$ year, then:

$$F = 100 \left[e^{0.06 \times 1/2} - 1\right] / \left[e^{0.06} - 1\right] = 100[0.03045/0.0618] = 100 \times 0.4925 = \$49.25,$$

which is **wrong**. This example intentionally used a low interest rate and short time period to indicate that *the length of the compounding period and the nominal interest rate period must be the same, which is one month and the corresponding interest is 0.005% per month.*

8.5.3 UNIFORM SERIES CONTINUOUS INTEREST SINKING FUND $(A/F, r, n)$ EXAMPLE

Lady Hillary has an option to buy a business in five years for $10,000.000. Her account pays a nominal interest rate of 12% per year. She plans to deposit money monthly in the account and wants to know what amount she must pay monthly to achieve the $10,000,000 at the end of 5 years. For monthly deposits, the interest must be compounded monthly and $r = 1\%$ per month.

$$(A/F, r, n) = [e^r - 1] / [e^{rn} - 1]. \tag{8.21}$$

Therefore,

$$
\begin{aligned}
A &= F\,[e^r - 1] / [e^{rn} - 1] && \text{(8.25)} \\
&= \$10,000,000 \left[e^{.01} - 1\right] / \left[e^{.01 \times 60} - 1\right] \\
&= \$10,000,000[1.010050 - 1] / \left[e^{.01 \times 60} - 1\right] \\
&= \$10,000,000[1.010050 - 1]/[1.8221188 - 1] \\
&= \$10,000,000[0.0122247] \\
&= \$122,247.
\end{aligned}
$$

Thus, Lady Hillary must deposit $122,247 at the end of every month to have her $10,000,000. Note that if there was no interest, the monthly payments would be $166,666 instead of the $122,247 per month.

This example intentionally used a high interest rate and long time period but the length of the compounding period and the nominal interest rate period must be the same. Thus, the 5 years of monthly payments indicates 60 monthly payments and the nominal interest rate per month is 1%.

8.5.4 UNIFORM SERIES CONTINUOUS INTEREST PRESENT WORTH $(P/A, r, n)$ EXAMPLE

Ronnie won the $240,000,000 lottery which will pay $12,000,000 per year for 20 years or he can take a single payment now which is discounted at an annual nominal interest rate of 9% per year. What is the amount that he would receive as a single payment. These would be beginning of year payments, so he would receive the first payment at time zero and then 19 more end-of-year payments:

$$(P/A, r, n) = [e^{rn} - 1] / [e^{rn} (e^r - 1)]. \tag{8.22}$$

Therefore,

$$P = A \ [e^{rn} - 1] / [e^{rn} (e^r - 1)]. \tag{8.26}$$

For this problem a payment must be made at time zero and then the following 19 end-of-period payments (which represent the remaining 19 beginning-of-year payments) result in:

$$P = A + A \ [e^{rn} - 1] / [e^{rn} (e^r - 1)]$$
$$= A \ \{1 + [e^{rn} - 1] / [e^{rn} (e^r - 1)]\}$$

$$P = \$12,000,000 \ \{1 + [e^{0.09 \times 19} - 1] / [e^{0.09 \times 19} (e^{0.09} - 1)]\}$$
$$= \$12,000,000 \ \{1 + [5.52896 - 1]/[5.52896 \times (1.09417 - 1)]\}$$
$$= \$12,000,000 \{1 + 8.75568\}$$
$$= \$117,068,160.$$

Note that the present worth is less than half of the lottery amount when the interest rate is 9% compounded continuously.

8.5.5 UNIFORM SERIES CONTINUOUS INTEREST CAPITAL RECOVERY FACTOR $(A/P, r, n)$ EXAMPLE

Queen Nancy has decided to purchase the Island of Happiness in the Ocean of Calm Waters for 2 billion dollars over 20 years at a 6% annual nominal interest rate compounded continuously. What would her annual end-of-year payments be to repay the 2 billion loan?

The factor of $(A/P, r, n)$ is:

$$(A/P, r, n) = [e^{rn} (e^r - 1)] / [e^{rn} - 1]. \tag{8.23}$$

Therefore,

$$A = P \ [e^{rn} (e^r - 1)] / [e^{rn} - 1] \tag{8.27}$$
$$A = \$2,000,000,000 \ \{[e^{.06 \times 20} (e^{.06} - 1)] / [e^{0.06 \times 20} - 1]\}$$
$$= \$2,000,000,000 \ \{[3.320117(1.0618365 - 1)] / [3.3320117 - 1]\}$$
$$= \$2,000,000,000 \{0.088037472\}$$
$$= \$176,074,944.$$

The total of the 20 year payments of $176 million will be approximately 3.52 billion which implies that 1.52 billion is paid in interest over the life of the investment which is more than 3/4's of the original loan value.

8.6 SUMMARY

This chapter has used the mathematical relationships of Chapter 7 to develop the discrete interest expressions for the single cash flow present worth and future worth and the discrete interest uniform series expressions for future worth, sinking fund, present worth, and capital recovery. Example problems using each of the formulas were presented. The factors for the discrete interest expressions developed were then converted for the continuous interest expressions. Example problems were presented using the nominal interest and the need to use the nominal interest based on the compounding period used was emphasized. A summary of the formulas is presented in Table 8.2 at the end of this chapter.

Table 8.2: Discrete and continuous compounding factors of the basic economic expressions

Notation:					
P = Present Worth A = uniform end-of-period i = discrete interest rate per compounding period					
F = Future Worth n = number of compounding periods r = continuous interest rate per compounding period					
Compounding Factors					
Payment Type	**Factor Name**	**Find**	**Given**	**Symbol**	**Formula**
A. Single Payment	Present Worth	P	F	(P/F, i, n)	$(1 + i)^{-n}$
(discrete interest)	Future Worth	F	P	(F/P, i, n)	$(1 + i)^n$
B. Uniform Payment or	Sinking Fund	A	F	(A/F, i, n)	$i/[(1 + i)^n -1]$
Uniform Series	Capital Recovery	A	P	(A/P, i, n)	$[i(1 + i)^n]/[(1 + i)^n -1]$
(discrete interest)	Future Worth	F	A	(F/A, i, n)	$[(1 + i)^n -1]/i$
	Present Worth	P	A	(P/A, i, n)	$[(1 + i)^n -1]/[i[(1 + i)^n]$
C. Single Payment	Present Worth	P	F	(P/F, r, n)	e^{-rn}
(continuous interest)	Future Worth	F	P	(F/P, r, n)	e^{rn}
D. Uniform Payment or	Sinking Fund	A	F	(A/F, r, n)	$[(e^r - 1)/(e^{rn} - 1)]$
Uniform Series	Capital Recovery	A	P	(A/P, r, n)	$[e^{rn} (e^r - 1)/(e^{rn} - 1)]$
(continuous interest)	Future Worth	F	A	(F/A, r, n)	$[(e^{rn} - 1)/(e^r - 1)]$
	Present Worth	P	A	(P/A, r, n)	$[(e^{rn} - 1)/(e^{rn}(e^r - 1))]$
Note: $i = = e^r -1$ and $(1+i)^n = e^{rn}$					

8.7 REFERENCES

[1] Creese, Robert C. and Adithan, M., *Strategic Cost Analysis for Project Managers and Engineers*, New Academic Science Limited, Tunbridge Wells, UK, pp. 39–48, 2012. 93, 111

[2] Park, Chan S., *Contemporary Engineering Economics*, 2nd ed., Addison-Wesley, Menlo Park, CA, p. 803, 1997. 93

[3] Newnan, Donald G., Eschenbach, Ted G., and Lavelle, Jerome P., *Engineering Economic Analysis*, 11th ed., Oxford University Press, New York, p. 655, 2012. 93

[4] Mehta, Merwan B., *Applied Engineering Economics Using Excel*, Industrial Press, Inc., South Norwalk, CT, p. 260, 2016. 93

[5] Whitman, David L. and Terry, Ronald E., *Fundamentals of Engineering Economics and Decision Analysis*, Morgan & Claypool Publishers, San Rafael, CA, p. 219, 2012. 93, 111

8.8 EVALUATIVE QUESTIONS

1. If the time period is 5 years and the interest rate is 10%, calculate the following values:

   (a) $(P/A, i, = 10\%, n = 5)$

   (b) $(F/A, i = 10\%, n = 5)$

   (c) $(A/P, i = 10\%, n = 5)$

   (d) $(A/F, i = 10\%, n = 5)$

   (e) $(P/F, i = 10\%, n = 5)$

   (f) $(F/P, i = 10\%, n = 5)$

2. If the time period is 5 years and the interest rate is 7.5%, calculate the following values:

   (a) $(P/A, i, = 7.5\%, n = 5)$

   (b) $(F/A, i = 7.5\%, n = 5)$

   (c) $(A/P, i = 7.5\%, n = 5)$

   (d) $(A/F, i = 7.5\%, n = 5)$

   (e) $(P/F, i = 7.5\%, n = 5)$

   (f) $(F/P, i = 7.5\%, n = 5)$

3. Engineer Jimmy wants to retire in 20 years and would like to have 1 million Euros at that time. If the interest rate is expected to be 5% over the next 20 years, what annual amount would he need to save?

4. Rosalynn has purchased a house with a loan of $500,000 Kroner. If the loan interest rate is 5%, what will be her annual payments over the life of the 20-year loan?

5. Gerald has decided to save for a trip in two years. If the monthly interest rate is 1% and he saves Rs. 2,500/month, how much will he have saved after 2 years?

6. Yi has decided to purchase a new moped and the discrete interest rate is 1/2% per month and the purchase price is 20,000 yuan. The payments will be at the end of the month; what is her expected monthly payment over the 3-year period?

7. Vladimir has purchased a new car and the discrete interest rate is 2% per month and the purchase price is 100,000 rubles.

   (a) What is the expected monthly payment over a 4-year period?

   (b) What is the total interest paid over the 4-year period?

8. Marlene won the Irish Sweepstakes of 20 million Irish pounds. The prize is actually 2 million pounds per year for 10 years.

   (a) If the payments are 10 beginning-of year payments and the discrete interest rate is 5%, what is the equivalent total amount she would receive if she took a single payment? Note: since payments are end-of-year, she has 9 end-of-year payments plus the initial first payment.

   (b) If the payments are 10 end-of-year payments and the discrete interest rate is 5%, what is the equivalent amount she would receive each year if she converts the payments to beginning-of-year payments.

9. If the time period is 5 years and the nominal continuous interest rate is 10%, calculate the values of the following factors:

   (a) $(P/A, r, = 10\%, n = 5)$

   (b) $(F/A, r = 10\%, n = 5)$

   (c) $(A/P, r = 10\%, n = 5)$

   (d) $(A/F, r = 10\%, n = 5)$

   (e) $(P/F, r = 10\%, n = 5)$

   (f) $(F/P, r = 10\%, n = 5)$

   Compare these values with the values in Problem 1.

10. If the time period is 3 months and the nominal continuous annual interest rate is 12%, calculate the values of the following factors:

    (a) $(P/A, r = ?, n = 3)$

(b) $(F/A, r =?, n = 3)$

(c) $(A/P, r =?, n = 3)$

(d) $(A/F, r =?, n = 3)$

(e) $(P/F, r =?, n = 3)$

(f) $(F/P, r =?, n = 3)$

11. Construct a table in a spreadsheet and calculate the expressions for the factors of $(P/F, i, n)$, $(F/P, i, n)$, $(P/A, i, n)$, $(A/P, i, n)$, $(F/A, i, n)$, and $(A/F, i, n)$ for $i = 10\%$ and $n = 1$–60 and $n = 100$. Compare the values calculated with those in the various reference books [1–5].

12. Construct a table in a spread sheet and calculate the expressions for the factors of $(P/F, r, n)$, $(F/P, r, n)$, $(P/A, r, n)$, $(A/P, r, n)$, $(F/A, r, n)$, and $(A/F, r, n)$ for $r = 10\%$ and $n = 1$–60 and $n = 100$. Compare the values calculated with those you calculated for Problem 9.

CHAPTER 9

Gradient Economic Factors and Equations

9.1 INTRODUCTION

Gradient expressions are more complex than the basic expressions in the previous chapter. The two major classifications of the gradient expressions are the uniform gradient and the geometric gradient. The uniform gradient is presented in two versions; the standard uniform gradient which starts in second period and the uniform ramp gradient which starts in the first period and appears like a ramp or step function. Similarly, the geometric gradient is presented in two versions; the standard geometric gradient in which the gradient does not start until the second period and the escalation gradient in which the gradient part starts in the first period. These gradients can be expressed with discrete or continuous interest expressions. Thus, the four gradient expressions will be developed for both the discrete and continuous interest and they are: the standard uniform gradient, the uniform ramp gradient, the geometric gradient, and the escalation gradient. Each of these will be initially described and the details of their derivation will be presented for the discrete interest case and then the expressions will be converted to the continuous interest case for the four systems. Some of these materials were presented in a previous work [1] and the next two references [2, 3] have some of the gradient expressions and present many more examples and problems, but some expressions developed are entirely new. Reference [4] is at a graduate level and uses a different approach by using Z-Transforms for some of the expressions developed. The uniform ramp gradient is not presented in most books and the escalation gradient is frequently not considered, but these expressions are quite useful when using the end-of period payments for annual increases starting in the first year.

9.2 STANDARD UNIFORM GRADIENT DISCRETE INTEREST

The standard uniform gradient can be expressed as a fixed amount which increases by the same amount in each of the following periods. The constant increase is called a uniform gradient appears to look somewhat like a ramp function which is delayed by one period. Most authors indicate that the uniform gradient first occurs at the end of the second period as indicated by Figure 9.1. Thus, there is one less payment than there are periods as the first gradient payment does not start until period 2; thus, if there is a gradient over a period of ten years, there will only be nine payments. The future worth of the gradient can be expressed by Equation (9.1) which starts with last term and proceeds back to the first term. The amount of the gradient is G, the interest rate is i, and the number of periods is n and a sketch is shown in Figure 9.1:

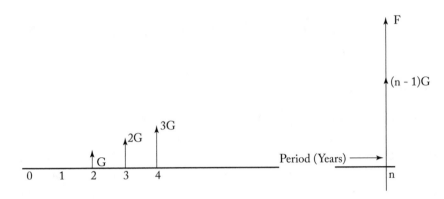

Figure 9.1: Standard uniform gradient for future worth derivation.

$$F = (n-1)G + (n-2)G(1+i) + (n-3)G(1+i)^2 + \cdots$$
$$\cdots + 2G(1+i)^{(n-3)} + G(1+i)^{(n-2)} + 0 + 0 \tag{9.1}$$
$$(\text{year } n) + (\text{year } n-1) + \cdots + (\text{year } 2) + (\text{year } 1) + (\text{time } 0).$$

If one multiplies Equation (9.1) by $(1+i)$ and then subtracts Equation (9.1) from that, one obtains Equation (9.2). Rearranging the terms and using the geometric series, the expression then becomes Equation (9.3):

$$(1+i)F = (n-1)G(1+i) + (n-2)G(1+i)^2 + (n-3)G(1+i)^3 + \cdots$$
$$\cdots + 2G(1+i)^{(n-2)} + G(1+i)^{(n-1)}$$
$$F = (n-1)G + (n-2)G(1+i) + (n-3)G(1+i)^2 + (n-4)G(1+i)^3 + \cdots$$
$$\cdots + G(1+i)^{(n-2)}$$

$$iF = -(n-1)G + G(1+i) + G(1+i)^2 + G(1+i)^3$$
$$+ G(1+i)^{(n-2)} + G(1+i)^{(n-1)} \tag{9.2}$$
$$= -nG + G \left[1 + (1+i) + (1+i)^2 + (1+i)^3 + \cdots \right.$$
$$\left. \cdots + (1+i)^{(n-3)} + (1+i)^{(n-2)} + (1+i)^{(n-1)} \right]$$
$$[\quad \text{this is the geometric series} \quad]$$
$$= -nG + G \left[(1+i)^n - 1 \right] / \left[(1+i) - 1 \right]$$
$$= -nG + G \left[(1+i)^n - 1 \right] / [i]$$
$$= G \{ [(1+i)^n - 1 - ni] / [i] \}. \tag{9.3}$$

Solving for F one obtains the uniform gradient discrete interest expression for the future worth becomes:

$$F = G \{ [(1+i)^n - 1 - ni] / [i^2] \} \tag{9.4}$$
$$= G(F/G, i, n).$$

The formula, represented by the factor $(F/G, i, n)$, is used to convert a discrete interest standard uniform gradient of G to a future worth F and this is what appears in Table 9.1 at the end of this section:

$$(F/G, i, n) = \{ [(1+i)^n - 1 - ni] / [i^2] \}. \tag{9.5}$$

The present worth of the uniform gradient discrete interest can be obtained by:

$$P = F/(1+i)^n = G \{ [(1+i)^n - 1 - ni] / [i^2 (1+i)^n] \}$$
$$= G \{ [(1+i)^n - 1 - ni] / [i^2 (1+i)^n] \} \tag{9.6}$$
$$= G(P/G, i, n).$$

Thus, the conversion formula to convert a uniform gradient discrete interest to a present worth is:

$$(P/G, i, n) = \{ [(1+i)^n - 1 - ni] / [i^2 (1+i)^n] \}. \tag{9.7}$$

The uniform series of the standard uniform gradient discrete interest can be obtained by

$$A = F \{ i / [(1+i)^n - 1] \} = G \{ [(1+i)^n - 1 - ni] / [i^2] \} \times (i / [(1+i)^n - 1])$$
$$= G [(1+i)^n - 1 - ni] / [i ((1+i)^n - 1)] \tag{9.8}$$
$$= G (A/G, i, n).$$

Thus, the conversion formula to convert a standard uniform gradient discrete interest to a uniform series is:

$$(A/G, i, n) = [(1+i)^n - 1 - ni] / [i ((1+i)^n - 1)]. \tag{9.9}$$

9.2.1 STANDARD UNIFORM GRADIENT DISCRETE INTEREST EXAMPLE

What would be the value of a standard uniform gradient of $200 per year for a period of 10 years. The first payment would be at the end of the second year and the last payment at the end of the 10th year. The interest rate is 5% .

(a) What is the final gradient payment?

Payment at 10 year $= (n - 1) \times \$200 = \$1,800$.

(b) What is the total payment of the gradient amounts, not including the interest?

Total Gradient Payments Made $= [n(n + 1)/2] \times \$200 = (9 \times 10/2) \times \$200 = \$9,000$.

(Payments occur only in the last 9 periods, so $n = 9$.)

(c) What is the total value including the compounding of interest at the end of year 10?

Total Value is Future Worth if found by using Equation (9.4) which is:

$$F = G \left\{ [(1 + i)^n - 1 - ni] / [i^2] \right\} = G \left\{ [(1 + i)^n - 1] / i^2 - n/i \right\}$$
$$= \$200 \left[(1.05)^{10} - 1 - 10 \times 0.05 \right] / (0.05)^2$$
$$= \$200[1.62889 - 1 - 0.50]/(0.0025)$$
$$= \$200[.12889]/0.0025$$
$$= \$10,311.$$

Note the effect of compounding interest results in a total interest gain of $1,311.

The present worth of the gradient can be found directly by Equation (9.6), which is:

$$P = G \left\{ [(1 + i)^n - 1 - ni] / [i^2(1 + i)^n] \right\}$$
$$= \$200 \left[(1.05)^{10} - 1 - 10 \times 0.05 \right] / \left[(0.05)^2 \times (1.05)^{10} \right]$$
$$= \$200[.12889]/[0.0040722]$$
$$= \$6,330.$$

The equivalent annual uniform series payment A can be found by Equation (9.9), which is:

$$A = G \left[(1 + i)^n - 1 - ni \right] / \left[i \left((1 + i)^n - 1 \right) \right]$$
$$= \$200 \left[(1.05)^{10} - 1 - 10 \times 0.05 \right] / \left[(0.05) \times \left((1.05)^{10} - 1 \right) \right]$$
$$= \$200[0.12889]/[0.031444]$$
$$= \$819.8.$$

Thus, it takes $819.8 uniform series payment to be equivalent to a $200 standard uniform gradient over a 10-year period at 5% interest. The equivalent uniform series payments will vary considerably as the time increment changes.

The payments have been and usually are considered end-of-period payments. These payments can be converted to beginning-of-period payments by dividing by the annual uniform series payment A by $(1 + i)$. Thus, the beginning-of-period payments are lower and for the previous end-of-period payment of $819.8 would be $780.7 as the interest rate is 5%.

9.3   UNIFORM RAMP GRADIENT DISCRETE INTEREST

The uniform ramp gradient starts in the first period and has the appearance of a ramp starting at zero. The future worth of the gradient can be expressed by Equation (9.10) which starts with last term and proceeds back to the first term. The amount of the gradient is G, the interest rate is i, and the number of periods is n. The subscript R is used to distinguish between the standard uniform gradient and the Uniform Ramp Gradient, as shown in Figure 9.2:

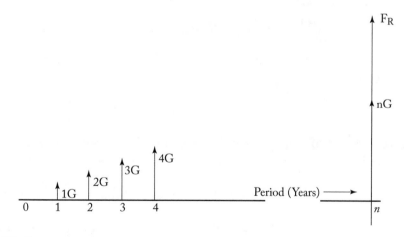

Figure 9.2: Uniform ramp gradient for future worth derivation.

$$F_R = (n)G + (n - 1)G(1 + i) + (n - 2)G(1 + i)^2 + \cdots$$
$$\cdots + 2G(1 + i)^{(n-2)} + G(1 + i)^{(n-1)} \tag{9.10}$$
$$(\text{year } n) + (\text{year } n - 1) + \cdots + (\text{year } 1).$$

If one multiplies Equation (9.10) by $(1 + i)$ and then subtracts Equation (9.10) from that, one obtains Equation (9.2). Rearranging the terms and using the geometric series, the expression then becomes Equation (9.3):

$$(1+i)F_R = nG\,(1+i)(n-1)G(1+i)^2 + (n-2)G(1+i)^3 + (n-3)G(1+i)^4 + \cdots$$
$$\cdots + 2G\,(1+i)^{(n-1)} + G(1+i)^n$$
$$F_R = (n)G + (n-1)G(1+i) + (n-2)G(1+i)^2 + (n-3)G(1+i)^3 + \cdots$$
$$\cdots + G(1+i)^{(n-1)}$$

$$\overline{}$$

$$iF_R = -(n)G + G(1+i) + G(1+i)^2 + G(1+i)^3 + \cdots$$
$$\cdots + G\,(1+i)^{(n-1)} + G\,(1+i)^n \tag{9.11}$$
$$= -nG + (1+i)G\,\left[1 + (1+i) + (1+i)^2 + (1+i)^3 + \cdots\right.$$
$$\left.\cdots + (1+i)^{(n-2)} + (1+i)^{(n-1)}\right]$$

$$[\quad \text{this part is the geometric series}\quad]$$
$$= -nG + (1+i)G\,[(1+i)^n - 1]\,/\,[(1+i) - 1]$$
$$= -nG + (1+i)G\,[(1+i)^n - 1]\,/\,[i]$$
$$= G\,\{[(1+i)^{n+1} - 1 - i(n+1)]\,/\,[i]\}\,. \tag{9.12}$$

Solving for F one obtains the uniform ramp gradient discrete interest expression for the future worth becomes:

$$F_R = G\,\{[(1+i)^{n+1} - 1 - i(1+n)]\,/\,[i^2]\}\,. \tag{9.13}$$

Note that the uniform ramp gradient is similar to the standard uniform gradient with n in the stardard gradient replaced by $(n+1)$ in the uniform ramp gradient:

$$F = G\,\{[(1+i)^n - 1 - ni]\,/\,[i^2]\}\,. \tag{9.6}$$

The formula, represented by the factor $(F_R/G, i, n)$, is used to convert a standard uniform gradient discrete interest of G to a future worth F and this is what appears in Table 9.1:

$$(F_R/G, i, n) = \{[(1+i)^{n+1} - 1 - i(1+n)]\,/\,[i^2]\}\,. \tag{9.14}$$

The value for the present worth of the uniform ramp gradient discrete interest can be obtained by converting the future worth to the present worth by:

$$P_R = F_R/(1+i)^n = G\,\{[(1+i)^{n+1} - 1 - i(1+n)]\,/\,[i^2]\}\,(1/(1+i)^n)$$
$$= G\,[(1+i)^{n+1} - 1 - i(1+n)]\,/\,[i^2(1+i)^n] \tag{9.15}$$
$$= G\,(P_R/G, i, n)\,.$$

Thus, the conversion formula to convert a uniform ramp gradient discrete interest to a present worth is:

$$(P_R/G, i, n) = \left\{ \left[(1+i)^{n+1} - 1 - i(1+n) \right] / \left[i^2 (1+i)^n \right] \right\}. \tag{9.16}$$

The value for the uniform series of the uniform ramp gradient discrete interest can be obtained by

$$\begin{aligned}
A_R &= F_R i / \left[(1+i)^n - 1 \right] = G \ \left\{ \left[(1+i)^{n+1} - 1 - i(1+n) \right] / \left[i^2 \right] \right\} \times i / \left[(1+i)^n - 1 \right] \\
&= G \ \left[(1+i)^{n+1} - 1 - i(1+n) \right] / \left[i \left((1+i)^n - 1 \right) \right] \\
&= G \ (A_R/G, i, n).
\end{aligned} \tag{9.17}$$

Thus, the conversion formula to convert a uniform ramp gradient discrete interest to a uniform series is:

$$(A_R/G, i, n) = \left[(1+i)^{n+1} - 1 - i(1+n) \right] / \left[i \left((1+i)^n - 1 \right) \right]. \tag{9.18}$$

9.3.1   UNIFORM RAMP GRADIENT DISCRETE INTEREST EXAMPLE

What would be the value of a uniform ramp gradient of $200 per year for a period of 10 years. The first payment would be at the end of the 1st year and the last payment at the end of the 10th year. The interest rate is 5%.

(a) What is the final payment?

Payment at year $10 = (n) \times \$200 = \$2,000$.

(b) What is the total payment of the gradient, not including the interest?

Total Payments Made $= [n(n+1)/2] \times \$200 = (10 \times 11/2) \times \$200 = \$11,000$.

(The payments occur in all 10 periods. The total payments are $2,000 more than the standard uniform gradient, that is: 10 years $\times$ $200/year = $2,000.)

(c) What is the total value including the compounding of interest at the end of year 10?

Total Value is Future Worth if found by using Equation (9.13) which is:

$$\begin{aligned}
F_R &= G \ \left\{ \left[(1+i)^{n+1} - 1 - i(1+n) \right] / \left[i^2 \right] \right\} \\
&= \$200 \left[(1.05)^{11} - 1 - 11 \times 0.05 \right] / (0.05)^2 \\
&= \$200 [1.71034 - 1 - 0.55] / (0.0025) \\
&= \$200 [.16034] / 0.0025 \\
&= \$12,827.
\end{aligned}$$

Note the effect of compounding interest results in a total interest gain of $1,827.

The present worth of the gradient can be found directly by Equation (9.15), which is:

$$\begin{aligned} P_R &= G\left[(1+i)^{n+1} - 1 - i(1+n)\right] / \left[i^2(1+i)^n\right] \\ &= \$200\left[(1.05)^{11} - 1 - 11 \times 0.05\right] / \left[(0.05)^2 \times (1.05)^{10}\right] \\ &= \$200[.16034]/[0.0040722] \\ &= \$7,875. \end{aligned}$$

The equivalent annual uniform series payment A can be found by Equation (9.17), which is:

$$\begin{aligned} A_R &= G\left[(1+i)^{n+1} - 1 - i(1+n)\right] / \left[i\left((1+i)^n - 1\right)\right] \\ &= \$200\left[(1.05)^{11} - 1 - 11 \times 0.05\right] / \left[(0.05) \times \left((1.05)^{10} - 1\right)\right] \\ &= \$200[0.16034]/[0.031444] \\ &= \$1019.8. \end{aligned}$$

Thus, it takes $1019.8 uniform series payment to be equivalent to a $200 uniform ramp gradient over a 10-year period. The equivalent uniform series payments will vary considerably as the time increment changes. Note that the amount of the uniform series payment equivalent for the uniform ramp gradient was $200 more than that for the standard uniform gradient for this problem. Thus, $A_R = A$ (Gradient) $+ A$ which is how the problem was solved previously by solving the standard gradient and then adding the additional amount.

9.4 GEOMETRIC GRADIENT DISCRETE INTEREST

The geometric gradient does not start until the second period. The present worth of the geometric gradient can be expressed by Equation (9.19). The initial amount is A_1 starts in period 1 and has the gradient amount g applied each of the follow periods up to the final period n. The present worth is obtained by discounting the gradient payments back to time zero. Figure 9.3 is a sketch of the payments for geometric gradient for deriving the present worth factor. Additional expressions must be developed when $g = i$ and the initial expressions presented are for $g \neq i$:

$$\begin{aligned} P &= A_1/(1+i) + A_1\left[(1+g)/(1+i)^2\right] + A_1\left[(1+g)^2/(1+i)^3\right] \\ &\quad + A_1\left[(1+g)^3/(1+i)^4\right] + \cdots + A_1\left[(1+g)^{(n-1)}/(1+i)^n\right]. \end{aligned} \tag{9.19}$$

Rearranging the terms to obtain the value of 1 for the first term of the geometric gradient and using the geometric series with the ratio of $(1+g)/(1+i)$, this equation can be reduced

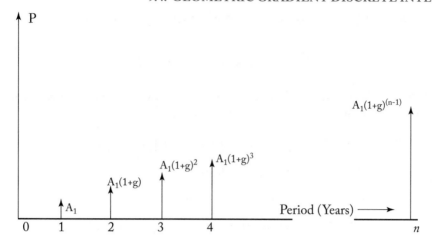

Figure 9.3: Standard geometric gradient for present worth derivation.

to:

$$P = [A_1/(1+i)]\left[1 + (1+g)/(1+i) + \{(1+g)/(1+i)\}^2 + \cdots + \{(1+g)/(1+i)\}^n\right]$$
$$= [A_1/(1+i)] \quad [\qquad\text{geometric series}\qquad]$$
$$= [A_1/(1+i)]\{[(1+g)/(1+i)]^n - 1\}/[(1+g)/(1+i) - 1]$$
$$= [A_1/(1+i)]\{[(1+g)/(1+i)]^n - 1\}/[[(1+g)-(1+i)]/(1+i)]$$
$$= A_1\{[(1+g)/(1+i)]^n - 1/[(g-i)]\}$$
$$P = A_1\{[1 - (1+g)/(1+i)]^n/[(i-g)]\}. \tag{9.20}$$

Thus, the geometric gradient present worth factor when $i \neq g$ is:

$$(P/A_1, g, i, n) = \{[1 - (1+g)/(1+i)]^n]/[(i-g)]\}. \tag{9.21}$$

If $i = g$, the denominator of Equation (9.21) would be zero which is a problem, so when $i = g$, Equation (9.19) can be arranged so that the ratio $(1+g)/(1+i)$ becomes 1 and thus:

$$P = A_1/(1+i) + A_1\left[(1+g)/(1+i)^2\right] + A_1\left[(1+g)^2/(1+i)^3\right]$$
$$+ A_1\left[(1+g)^3/(1+i)^4\right] + \cdots + A_1\left[(1+g)^{(n-1)}/(1+i)^n\right] \tag{9.19}$$
$$= [A_1/(1+i)]\left[1 + (1+g)/(1+i) + \{(1+g)/(1+i)\}^2 + \cdots\right.$$
$$\left.\cdots + \{(1+g)/(1+i)\}^{(n-1)}\right]$$
$$= [A_1/(1+i)][1 + 1 + 1 + \cdots + 1]$$
$$P = A_1/(1+i)[n]$$
$$P = nA_1/(1+i). \tag{9.22}$$

Thus, one has a rather simple expression when $i = g$ and the geometric gradient present worth factor expression becomes:

$$(P/A_1, i = g, n) = \{n/(1 + i)\}. \tag{9.23}$$

The value for the future worth of the geometric gradient can be obtained by multiplying by $(1 + i)^n$ to the present worth of the geometric gradient to obtain:

$$F = P(1 + i)^n = A_1 \{[1 - (1 + g)/(1 + i)^n]/[(i - g)]\} \times (1 + i)^n$$
$$F = A_1 \{[(1 + i)^n - (1 + g)^n]/[i - g]\}. \tag{9.24}$$

Thus, the geometric gradient future worth factor is:

$$(F/A_1, g, i, n) = \{[(1 + i)^n - (1 + g)^n]/[i - g]\}. \tag{9.25}$$

If $i = g$, then the future worth factor for the geometric gradient is the present worth factor for the geometric gradient multiplied by $(1 + i)^n$ which is:

$$F = P(1 + i)^n = [(nA_1/(1 + i))] \times (1 + i)^n$$
$$F = [nA_1(1 + i)^{n-1}]. \tag{9.26}$$

Thus, a simple expression occurs when $i = g$ and the geometric gradient future worth factor expression becomes:

$$(F/A_1, i = g, n) = \{n(1 + i)^{n-1}\}. \tag{9.27}$$

The value for the uniform series of the geometric gradient can be obtained by:

$$A = A_1 (F/A_1 g, i, n) \times (A/F, i, n)$$
$$A = A_1 \{[(1 + i)^n - (1 + g)^n]/[i - g]\} \times \{i/[(1 + i)^n - 1]\}. \tag{9.28}$$

Thus, the geometric gradient uniform series factor is:

$$(A/A_1, g, i, n) = \{[(1 + i)^n - (1 + g)^n]/[i - g]\} \times \{i/[(1 + i)^n - 1]\}. \tag{9.29}$$

If $i = g$, then the uniform series factor for the geometric gradient is the future worth factor for the geometric gradient multiplied by $[i/[(1 + i)^n - 1]]$ which is:

$$A = F[i/[(1 + i)^n - 1]] = (nA_1(1 + i)^{n-1})[i/[(1 + i)^n - 1]]$$
$$A = [niA_1(1 + i)^{n-1}]/[(1 + i)^n - 1]. \tag{9.30}$$

Finally, the geometric gradient uniform series factor expression when $i = g$ becomes:

$$(A/A_1, i = g, n) = [ni(1 + i)^{n-1}]/[(1 + i)^n - 1]. \tag{9.31}$$

9.4.1   GEOMETRIC GRADIENT DISCRETE INTEREST EXAMPLE

What would be the value at the end of 10 years of a geometric gradient of 10% if the initial amount was $200 for a period of 10 years. The first payment would be at the end of the first year and the last payment at the end of the 10th year. The interest rate is 5%.

(a) What is the final gradient payment? $200(1 + 0.10)^9 = \$100(2.3574) = \$471.5$.

(b) What is the total payments (TP) of the gradients not including the interest?

$$
\begin{aligned}
TP &= A_1 \left(1 + (1 + g) + (1 + g)^2 + \cdots + (1 + g)^{n-1}\right) \\
&= A_1 \left((1 + g)^n - 1\right) / (1 + g - 1) \\
&= A_1 \left((1 + g)^n - 1\right) / g \qquad\qquad\qquad (9.32) \\
&= \$200 \left(1.1^{10} - 1\right) / 0.10 \\
TP &= \$3,187.
\end{aligned}
$$

(c) What is the future worth including the compounding of interest?

$$
\begin{aligned}
F &= A_1 \left\{ [(1 + i)^n - (1 + g)^n] / [i - g] \right\} \qquad\qquad (9.24) \\
&= \$200 \left\{ [(1 + 0.05)^{10} - (1 + 0.10)] \right\} / [0.05 - 0.10] \\
&= \$200[-.9648478]/[-.05] = \$100(19.2969) \\
F &= \$3,859.
\end{aligned}
$$

(d) Therefore, the total interest earned over the 10 years would be:

$$
F - TP = \$3,859 - \$3,187 = \$672.
$$

(e) What is the present worth of the gradient?

$$
\begin{aligned}
P &= A_1 \left\{ [1 - [(1 + g)/(1 + i)]^n] / [(i - g)] \right\} \qquad\qquad (9.20) \\
&= \$200 \left\{ [1 - [(1 + 0.10)/(1 + 0.05)]^{10}] / [(0.05 - 0.10)] \right\} \\
&= \$200[-0.59233]/[-.05] \\
P &= \$2,369.
\end{aligned}
$$

(f) What is the equivalent annual uniform series payment A?

$$A = A_1 \left\{ [(1 + i)^n - (1 + g)^n] / [i - g] \right\} \left\{ i / [(1 + i)^n - 1] \right\} \qquad (9.28)$$
$$= \$200 \left\{ [(1.05)^{10} - (1.10)^{10}] / (0.05 - 0.10) \times [0.05 / 1.05^{10} - 1] \right\}$$
$$= \$200 \{[(19.2970] \times [(0.079504]\}$$
$$= \$307.$$

Note that the geometric gradient usually is smaller than the uniform ramp gradient or the standard uniform gradient as it is a percentage gradient rather than a fixed amount.

9.5   ESCALATION GRADIENT DISCRETE INTEREST

Escalation is used in construction projects as the cost of materials and labor are expected to increase over time and this is done with the escalation rate. Since traditional engineering economy expressions use end-of-period payments, the escalation must start in the first period. The projects estimates are done long before the project starts and the inflation effects will need to be included starting with the first period. This is the difference between escalation and the geometric gradient series which is similar to the difference needed for the uniform ramp gradient series and the standard uniform gradient. The illustration of the escalation payments and present worth is below and the present worth of the escalation can be expressed by Equation (9.33). The symbol E will be used to indicate escalation, but is used in the same manner as g except it is also in the first period. The escalation gradient is illustrated in Figure 9.4.

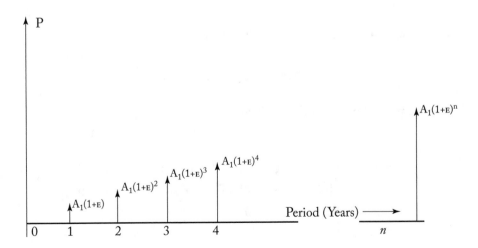

Figure 9.4: Escalation gradient for present worth derivation.

$$P_E = A_1(1 + E)/(1 + i) + A_1\left[(1 + E)^2/(1 + i)^2\right] + A_1\left[(1 + E)^3/(1 + i)^3\right]$$
$$+ A_1\left[(1 + E)^4/(1 + i)^4\right] + \cdots + A_1\left[(1 + E)^n/(1 + i)^n\right]$$
$$= [A_1(1 + E)/(1 + i)]\left[1 + (1 + E)/(1 + i) + \{(1 + E)/(1 + i)\}^2 + \cdots\right.$$
$$\cdots + \{(1 + E)/(1 + i)\}^n]$$
$$[\quad \text{Geometric Series} \quad]$$
$$= [A_1(1 + E)/(1 + i)]\left[[(1 + E)/(1 + i)]^n - 1\right]/[(1 + E)/(1 + i) - 1]$$
$$= [A_1(1 + E)/(1 + i)]\left[[(1 + E)/(1 + i)]^n - 1\right]/[[(1 + E) - (1 + i)]/(1 + i)]$$
$$= A_1\left\{[(1 + E)/(E - i)]\left[[(1 + E)/(1 + i)]^n - 1\right]\right\}$$
$$P_E = A_1\left\{[(1 + E)/(E - i)]\left[[(1 + E)/(1 + i)]^n - 1\right]\right\}. \tag{9.33}$$

Equation (9.33) is the expression for the escalation gradient and E is the escalation amount, and the symbol E is used instead of g. The escalation gradient present worth factor to obtain the present worth of an escalation gradient is.

$$(P_E/A_1, E, i, n) = \left\{[(1 + E)/(E - i)][(1 + E)/(1 + i)]^n - 1\right\}. \tag{9.34}$$

If $i = E$, Equations (9.33) and (9.34) will have denominators that are zero, so the equations must be derived initially using the case where $i = E$. This is solved in the same manner as was done for the standard geometric gradient:

$$P_E = A_1(1 + E)/(1 + i) + A_1\left[(1 + E)^2/(1 + i)^2\right] + A_1\left[(1 + E)^3/(1 + i)^3\right] + \cdots$$
$$\cdots + A_1\left[(1 + E)^n/(1 + i)^n\right]$$
$$= [A_1(1 + E)/(1 + i)]\left[1 + (1 + E)/(1 + i) + \{1 + E)/(1 + i)\}^2 + \cdots\right.$$
$$\cdots + \left\{(1 + E)^{(n-1)}/(1 + i)\right\}^{(n-1)}]$$
$$= [A_1 \times 1][1 + 1 + 1 \cdots + 1]$$
$$P_E = nA_1. \tag{9.35}$$

Thus, for this special case;

$$(P_E/A_1, i = E, n) = n. \tag{9.36}$$

The equation for the future worth of the escalation gradient can be determined from the present worth equation by:

$$F_E = A_1 \times (P_E/A_1, E, i, n) \times (F/P, i.n)$$
$$F_E = A_1 \times \left\{[(1 + E)/(E - i)][(1 + E)/(1 + i)]^n - 1\right\} \times (1 + i)^n$$
$$F_E = A_1\left\{[(1 + E)/(E - i)][(1 + E)^n - (1 + i)^n]\right\}. \tag{9.37}$$

This also results in the factor being:

$$(F_E/A_1, E, i, n) = \{[(1 + E)/(E - i)] [(1 + E)^n - (1 + i)^n]\}. \tag{9.38}$$

For the special case where $i = E$, then

$$F_E = A_1 \times (P_E/A_1, i = E, n) \times (F/P, i, n)$$
$$= A_1 n (1 + i)^n \tag{9.39}$$

and thus

$$(F_E/A_1 i = e, n) = n(1 + i)^n. \tag{9.40}$$

The equation for the uniform series of the escalation gradient can be determined from the present worth equation by:

$$A_E = A_1 \times (P_E/A_1, E, i, n) \times (A/P, i.n)$$
$$A_E = A_1 \times \{[(1 + E)/(E - i)][(1 + E)/(1 + i)]^n - 1\} \times \{[i((1 + i)^n / [((1 + i)^n - 1)]\}$$
$$A_E = A_1 \{[i(1 + E)/(E - i)] [(1 + E)^n - (1 + i)^n] / [(1 + i)^n - 1]\}. \tag{9.41}$$

This also results in the factor

$$(A_E/A_1, E, i, n) = \{[i(1 + E)/(E - i)] [(1 + E)^n - (1 + i)^n] / [(1 + i)^n - 1]\}. \tag{9.42}$$

For the special case where $i = E$, then

$$A_E = A_1 \times (P/A_1, i = E, n) \times (A/P, i.n)$$
$$= A_1 \{ni(1 + i)^n / [((1 + i)^n - 1)]\}. \tag{9.43}$$

And thus,

$$(A_E/A_1 i = E, n) = \{ni(1 + i)^n / [((1 + i)^n - 1)]\}. \tag{9.44}$$

9.5.1 ESCALATION GRADIENT DISCRETE INTEREST EXAMPLE

What would be the value at the end of 10 years of an escalation gradient of 10% if the initial amount was $200 for a period of 10 years. The first payment would be at the end of the first year and the last payment at the end of the 10th year. The interest rate is 5%.

(a) What is the final payment? $200(1 + 0.10)^{10} = \$200(2.5937) = \$518.8 = \$519$.

(b) What is the total payments (TP) of the gradients not including the interest?

$$TP_E = A_1 ((1 + E) + (1 + E)^2 + \cdots + (1 + E)^n)$$
$$= A_1(1 + E) (1 + (1 + E)^1 + \cdots + (1 + E)^{n-1})$$
$$= A_1(1 + E) ((1 + E)^n - 1) / (1 + E - 1)$$
$$= A_1(1 + E) ((1 + E)^n - 1) / E \tag{9.45}$$
$$= \$200(1.1) (1.1^{10} - 1) / 0.10$$
$$TP_E = \$3,506.$$

(c) What is the future worth including the compounding of interest?

$$F_E = A_1 \{[(1 + E)/(E - i)]\,[(1 + E)^n - (1 + i)^n]\} \tag{9.37}$$
$$= \$100[(1 + 0.10)/(0.10 - 0.05)]\,\{[(1 + 0.10)^{10} - (1 + 0.05)^{10}]\}$$
$$= \$200[22.0][0.964847]$$
$$F_E = \$4{,}245.$$

(d) Therefore, the total interest earned over the 10 years would be:

$$F_E - TP_E = \$4245 - \$3506 = \$739$$

(e) What is the present worth of the escalation gradient?

$$P_E = A_1 \{[(1 + E)/(E - i)]]\,[[(1 + E)/(1 + i)]^n - 1]\} \tag{9.33}$$
$$= \$200 \{[(1 + 0.10)/(0.10 - 0.05)] \times [[(1 + 0.10)/(1 + 0.5)]^{10} - 1]\}$$
$$= \$200 \times (1.1/0{,}05) \times [0.59233]$$
$$= \$2{,}606.$$

(f) What is the equivalent annual escalation payment A_e?

$$A_E = A_1 \{[i\,(1 + E)/(E - i)]\,[(1 + E)^n - (1 + i)^n]\,/\,[(1 + i)^n - 1]\} \tag{9.41}$$
$$= A_1 \{[i\,(1 + E)/(E - i)]\,[(1 + E)^n - (1 + i)^n]\,/\,[(1 + i)^n - 1]\}$$
$$= \$200[0.05(1 + 0.10)/(0.10 - 0.05)]\,\big[(1.10)^{10} - (1.05)^{10}\big]\,/\,\big[(1.05)^{10} - 1\big]$$
$$= \$200[1.1, 0.964847]/[0.62889]$$
$$A_E = \$338.$$

Note that the escalation gradient ($338) is larger than the geometric gradient ($307) as the gradient-escalation starts in the first period rather than the second period. This difference results in the 10% as the escalation gradient has one more escalation. A summary of all the discrete formulas is in Table 9.1. This difference is not only for A_E but also for P_E or F_E.

Table 9.1: Discrete compounding factors of economic expressions—discrete payments and discrete interest

Payment Type	Factor Name	Find	Given	Symbol	Formula
A. Single Payment	Present Worth	P	F	(P/F, i, n)	$(1+i)^{-n}$
	Future Worth (Compound Amount)	F	P	(F/P, i, n)	$(1+i)^n$
B. Uniform Payment (Uniform Series)	Sinking Fund	A	F	(A/F, i, n)	$i / [(1+i)^n - 1]$
	Capital Recovery	A	P	(A/P, i, n)	$[i(1+i)^n] / [(1+i)^n - 1]$
	Compound Amount	F	A	(F/A, i, n)	$[(1+i)^n - 1] / i$
	Present Worth	P	A	(P/A, i, n)	$[(1+i)^n - 1] / [i(1+i)^n]$
C. Uniform Gradient Expression					
Standard Uniform Gradient	Uniform Gradient Present Worth	P	G	(P/G, i, n)	$[((1+i)^n - 1 - ni) / (i^2 (1+i)^n)]$
	Uniform Gradient Future Worth	F	G	(F/G, i, n)	$[((1+i)^n - 1 - ni) / i^2]$
	Uniform Gradient Uniform Series	A	G	(A/G, i, n)	$[((1+i)^n - 1 - ni) / ((1+i)^n - 1)]$
Uniform Ramp Gradient	Uniform Ramp Gradient Present Worth	P_R	G	$(P_R/G, i, n)$	$[((1+i)^{n+1} - 1 - i(n+1)) / (i^2 (1+i)^n)]$
	Uniform Ramp Gradient Future Worth	F_R	G	$(F_R/G, i, n)$	$[((1+i)^{n+1} - 1 - i(n+1)) / (i^2)]$
	Uniform Ramp Gradient Uniform Series	A_R	G	$(A_R/G, i, n)$	$[((1+i)^{n+1} - 1 - i(n+1)) / (i(1+i)^n - 1)]$
D. Geometric Gradient Expression					
Geometric Gradient	Geometric Gradient Present Worth	P	A_1, g	$(P/A_1 g, i, n)$	$[(1-((1+g)^n / (1+i)^n))/(i-g)]$
			If g=i	$(P/A_1, g=i, n)$	$n/(1+i)$
	Geometric Gradient Future Worth	F	A_1, g	$(F/A_1, g, i, n)$	$[((1+i)^n - (1+g)^n)]/[i-g]$
			If g=i	$(F/A_1, g=i, n)$	$n(1+i)^{(n-1)}$
	Geometric Gradient Uniform Series	A	A_1, g	$(A/A_1, g, i, n)$	$[(i((1+i)^n - (1+g)^n))/((i-g)((1+i)^n-1))]$
			If g=i	$(A/A_1, g=i, n)$	$[ni(1+i)^{(n-1)}]/[(1+i)^n -1]$
Escalation Gradient	Escalation Gradient Present Worth	P_E	A_1, E	$(P_E/A_1, E, i, n)$	$[(1+E)/(E-i)][((1+E)/(1+i))^n -1]$
			If E=i	$(P_E/A_1, E=i, n)$	n
	Escalation Gradient Future Worth	F_E	A_1, E	$(F_E/A_1, E, i, n)$	$[(1+E)/(E-i)][((1+E)^n - (1+i)^n]$
			If E=i	$(F_E/A_1, E=i, n)$	$n(1+i)^n$
	Escalation Gradient Uniform Series	A_E	A_1, E	$(A_E/A_1, E, i, n)$	$[(i(1+E)/(E-i)*(1+E)^n - (1+i)^n)]/[(1+i)^n -1]$
			If E=i	$(A_E/A_1, E=i,n)$	$ni(1+i)^n /[(1+i)^n -1]$

Notation:
P=Present Worth;   i = effective discrete interest rate per period;   A=uniform end-of-period payments;   n = number of periods;
F=Future Worth;   g=Geometric Gradient Rate;   G=Uniform Gradient Amount;   E = Escalation Gradient Rate;
A_1 = Initial Geometric Gradient Amount and Initial Escalation Gradient Amount

9.6 STANDARD UNIFORM GRADIENT CONTINUOUS INTEREST FORMULAS

The basic relationships between the nominal (r) and market (i) interest will be used to convert the discrete interest formulas to continuous interest formulas similar to that done in Chapter 8. Most books have only small discussions of continuous interest, but Park and Sharp-Bette is a great reference [4]. The relationships used are:

$$i = e^r - 1 \tag{9.46}$$

and

$$(1 + i) = e^r \tag{9.47}$$

and similarly

$$(1 + i)^n = e^{rn} \tag{9.48}$$

and

$$r = \ln(1 + i). \tag{9.49}$$

The factors for the standard uniform gradient discrete interest were:

$$(F/G, i, n) = \left\{ [(1 + i)^n - 1 - ni] / [i^2] \right\} \tag{9.5}$$
$$(P/G, i, n) = \left\{ [(1 + i)^n - 1 - ni] / [i^2(1 + i)^n] \right\} \tag{9.7}$$

and

$$(A/G, i, n) = \left\{ [(1 + i)^n - 1 - ni] / [i \, ((1 + i)^n - 1)] \right\}. \tag{9.9}$$

The conversion of these factors for the standard uniform gradient continuous interest factors result in the following equations:

$$F = G(F/G, r, n) = G \left\{ [e^{rn} - 1 - n(e^r - 1)] / [(e^r - 1)]^2 \right\} \tag{9.50}$$

$$P = G(P/G, r, n) = G \left\{ [e^{rn} - 1 - n(e^r - 1)] / [(e^r - 1)]^2 \, e^{rn} \right\} \tag{9.51}$$

$$A = G(A/G, r, n) = G \left\{ [e^{rn} - 1 - n(e^r - 1)] / [(e^r - 1) \, (e^{rn} - 1)] \right\}. \tag{9.52}$$

9.6.1 STANDARD UNIFORM GRADIENT CONTINUOUS INTEREST EXAMPLE

The example will be the same as that in Section 9.2.1 to illustrate the difference between discrete and continuous. What would be the value of a standard uniform gradient (G) of $200 per year for a period of 10 years. The first payment would be at the end of the second year and the last payment at the end of the 10th year. The continuous interest rate is 5%.

(a) What is the final payment?

Payment at 10 year $= (n - 1) \times \$200 = \$1{,}800$.

(b) What is the total payment of the gradient, not including the interest?

Total Payments Made $= [n(n + 1)/2] \times \$100 = (9 \times 10/2) \times 200 = \$9{,}000$.

(Payments occur only in 9 periods even though there are 10 periods, so $n = 9$.)

(c) What is the total value including the continuous compounding of interest at the end of year 10?

Total Value is Future Worth of the standard uniform gradient continuous interest found by using Equation (9.50) which is:

$$F = G \ (F/G, r = 5\%, n = 10)$$
$$= G \ \left\{[e^{rn} - 1 - n \ (e^{r} - 1)] / [(e^{r} - 1)]^{2}\right\}$$
$$= \$200 \left[e^{.05 \times 10} - 1 - 10 \ (e^{0.05} - 1)\right] / \left[(e^{0.05} - 1)^{2}\right]$$
$$= \$200[1.64872 - 1 - 10(1.05127 - 1)] / \left[(1.05127 - 1)^{2}\right]$$
$$= \$200[.13601]/[0.0026287]$$
$$= \$200[51.74]$$
$$= \$10{,}348.$$

This is slightly greater than the discrete interest (10,312) as the equivalent discrete interest would be 5.13% instead of the 5.00% applied in the discrete interest calculations.

Note the effect of continuous interest results in a total interest gain of $1,348 which is slightly greater than the discrete case.

The present worth of the gradient can be found directly by the factor Equation (9.51), which is:

$$P = G(P/G, r = 5\%, n = 10)$$
$$= G \ \left\{[e^{rn} - 1 - n \ (e^{r} - 1)] / [(e^{r} - 1)]^{2} \ e^{rn}\right\}$$
$$= \$200 \left[e^{.05 \times 10} - 1 - 10 \ (e^{0.05} - 1)\right] / \left[(e^{0.05} - 1)^{2} \ e^{.05 \times 10}\right]$$
$$= \$200[1.64872 - 1 - 10(1.05127 - 1)] / \left[(1.05127 - 1)^{2} \times 1.64872\right]$$
$$= \$200[.13601]/[0.004334]$$
$$= \$6{,}276.$$

The discrete interest value of the present worth was \$6,330 and thus the difference is relatively small. The lower present worth is due to the slightly higher discount rate of the continuous compounding vs. discrete compounding.

The equivalent annual uniform series continuous interest payment can be found by Equation (9.52), which is:

$$
\begin{aligned}
A &= G\ (A/G, r = 5\%, n = 10) \\
&= G\ \{[e^{rn} - 1 - n\ (e^r - 1)]\ /\ [(e^r - 1)\ (e^{rn} - 1)]\} \\
&= \$200\left[e^{.05 \times 10} - 1 - 10\left(e^{0.05} - 1\right)\right]\ /\ \left[\left(e^{0.05} - 1\right)\left(e^{.05 \times 10} - 1\right)\right] \\
&= \$200[1.64872 - 1 - 10(1.05127 - 1)]/[(1.05127 - 1) \times (1.64872 - 1)] \\
&= \$200[.13601]/[0.0332606] \\
&= \$818.
\end{aligned}
$$

Thus, it takes \$818 uniform series continuous payment to be equivalent to a \$200 standard uniform gradient of a 10-year period. The equivalent uniform series payments will vary considerably as the time increment changes. The discrete interest and continuous interest values have only a slight difference as the interest rate is low and the compounding periods are relatively low.

9.7   RAMP UNIFORM GRADIENT CONTINUOUS INTEREST FORMULAS

The discrete interest formulas for uniform ramp gradient from Section 9.3 are presented and then converted into the continuous formulas for the uniform ramp gradient:

$$
\begin{aligned}
(F_R/G, i, n) &= \{[(1 + i)^{n+1} - 1 - i(1 + n)]\ /\ [i^2]\} & (9.13)\\
(P_R/G, i, n) &= \{[(1 + i)^{n+1} - 1 - i(1 + n)]\ /\ [i^2(1 + i)^n]\} & (9.16)\\
(A_R/G, i, n) &= [(1 + i)^{n+1} - 1 - i(1 + n)]\ /\ [i\ ((1 + i)^n - 1)] . & (9.18)
\end{aligned}
$$

The conversion of these factors from discrete to continuous interest factors results in the factor:

$$
(F_R/G, r, n) = \left\{\left[e^{r(n+1)} - 1 - (e^r - 1)(1 + n)\right]\ /\ \left[(e^r - 1)^2\right]\right\} \tag{9.53}
$$

$$
(P_R/G, r, n) = \left[e^{r(n+1)} - 1 - (e^r - 1)(1 + n)\right]\ /\ \left[(e^r - 1)^2\ e^{rn}\right] \tag{9.54}
$$

$$
(A_R/G, r, n) = \left[e^{r(n+1)} - 1 - (e^r - 1)(1 + n)\right]\ /\ [(e^r - 1)(e^{rn} - 1)] . \tag{9.55}
$$

9.7.1   UNIFORM RAMP GRADIENT CONTINUOUS INTEREST EXAMPLE

What would be the value of a uniform ramp gradient continuous interest of \$200 per year for a 10-year period with a nominal interest rate of 5%. The first payment would be at the end of the first year and the last payment at the end of the 10th year.

(a) What is the final payment?

Payment at year $10 = (n) \times \$200 = \$2,000.$

(b) What is the total payment of the gradient, not including the interest?

Total Payments Made $= [n(n+1)/2] \times \$200 = (11 \times 10/2) \times \$200 = \$11,000.$

(The payments occur in all 10 periods.)

(c) What is the total value including the compounding of the continuous interest at the end of year 10?

Total Value is Future Worth if found by using Equation (9.53) which is:

$$(F_R/G, r, n) = \left\{ \left[e^{r(n+1)} - 1 - (e^r - 1)(1 + n) \right] / \left[(e^r - 1)^2 \right] \right\} \qquad (9.53)$$

$$F_R = G \left\{ \left[e^{r(n+1)} - 1 - (e^r - 1)(1 + n) \right] / \left[(e^r - 1)^2 \right] \right\}$$

$$= \$200 \left[e^{0.05(10+1)} - 1 - (e^{.05} - 1)(1 + 10) \right] / \left(e^{0.05} - 1 \right)^2$$

$$= \$200[1.73325 - 1 - (1.051271 - 1)(11)]/(0.0025)$$

$$= \$200[.169269]/0.002629$$

$$= \$12,877.$$

Note the effect of compounding interest results in a total interest gain of \$1,877.

The present worth of the gradient can be found directly by using Equation (9.54), which is:

$$(P_R/G, r, n) = [(e^{r(n+1)} - 1 - (e^r - 1)(1 + n)]/[(e^r - 1)^2 e^{rn}] \qquad (9.54)$$

$$P_R = G[(e^{r(n+1)} - 1 - (e^r - 1)(1 + n)]/[(e^r - 1)^2 e^{rn}]$$

$$= \$200\{[(e^{0.05(10+1)} - 1 - (e^{0.05} - 1)(10 + 1)]/[(e^{0.05} - 1)^2 \times e^{0.05 \times 10}]$$

$$= \$200[0.16927]/[0.0043340]$$

$$= \$7,811.$$

The equivalent annual uniform series payment A can be found by Equation (9.55), which is:

$$(A_R/G, i, n) = \left[\left(e^{r(n+1)} - 1\right) - (e^r - 1)(1+n)\right] / [(e^r - 1)(e^{rn} - 1)] \qquad (9.55)$$

$$A_R = G \left[\left(e^{r(n+1)} - 1 - (e^r - 1)\right)(1+n)\right] / [(e^r - 1)(e^{rn} - 1)]$$

$$= \$200 \left[e^{0.05(10+1)} - 1 - (e^{0.05} - 1)(1+10)\right] /$$

$$/ \left[(e^{0.05} - 1)(e^{0.05 \times 10} - 1)\right]$$

$$= \$200[0.16927]/[0.0332607]$$

$$= \$1,018.$$

Thus, it takes $1,018 uniform series payment to be equivalent to a $200 uniform ramp gradient over a 10-year period. The equivalent uniform series payments will vary considerably as the time increment changes. Note that the amount of the uniform series payment equivalent for the uniform ramp gradient was $200 more than that for the standard uniform gradient for this problem which was expected.

9.8 GEOMETRIC GRADIENT CONTINUOUS INTEREST FORMULAS

The geometric gradient is often called the exponential gradient, and thus the geometric gradient continuous interest formulas may also be called the exponential gradient continuous interest formulas. The formulas from the discrete interest sections will be converted to continuous interest formulas. However, the formulas (9.46)–(9.49) did not include the conversion of g to a continuous expression. The formulas for $b = r$ will be left for student development.

Similarly, one would need relationships to convert the discrete gradient interest (g) to a continuous interest rate which will use the symbol (b). Therefore:

$$g = e^b - 1 \qquad (9.56)$$

and

$$(1 + g) = e^b \qquad (9.57)$$

and thus

$$(1 + g)^n = e^{bn} \qquad (9.58)$$

and

$$b = \ln(1 + g). \qquad (9.59)$$

The factors for the standard geometric gradient with discrete interest are:

$$(P/A_1, g, i, n) = \{[1 - (1 + g)/(1 + i)]^n]/[(i - g)]\} \tag{9.21}$$

$$(F/A_1, g, i, n) = \{[(1 + i)^n - (1 + g)^n]/[i - g]\} \tag{9.25}$$

$$(A/A_1, g, i, n) = \{[(1 + i)^n - (1 + g)^n]/[i - g]\}\{i/[(1 + i)^n - 1]\}. \tag{9.29}$$

The conversion of these factors the geometric gradient continuous interest result in the following expressions for the cases where $r \neq b$:

$$(P/A_1, b, r, n) = \left\{[1 - (e^{bn}/(e^{rn}))]/\left[\left(e^r - e^b\right)\right]\right\} \tag{9.60}$$

$$(F/A_1, b, r, n) = \left\{[(e^{rn} - e^{bn})]/\left[\left(e^r - e^b\right)\right]\right\} \tag{9.61}$$

$$(A/A_1, b, r, n) = \left\{[(e^{rn} - e^{bn})]/\left[\left(e^r - e^b\right)\right]\right\}\{(e^r - 1)/[e^{rn} - 1]\}. \tag{9.62}$$

9.8.1 GEOMETRIC GRADIENT CONTINUOUS INTEREST EXAMPLE

What would be the value at the end of 10 years of a geometric continuous gradient of $10\%(b = 10\%)$ if the initial amount was $200 for a period of 10 years. The first payment would be at the end of the 1st year and the last payment at the end of the 10th year. The interest rate is $5\%(r = 5\%)$. (Note: if $b = 10\%$, then $g = 10.52\%$.)

(a) What is the final payment? $200e^{0.10\times9} = \$200(e^{0.9}) = \$492$.

(b) What is the total payments (TP) of the gradients not including the interest?

$$TP = A_1\left(1 + (e^{0.1}) + (e^{0.1})^2 + \cdots + (e^{0.1})^{n-1}\right) = A_1\left((e^{bn} - 1))/\left(e^b - 1\right)\right.$$

$$= A_1\left(e^{bn} - 1\right)/\left(e^b - 1\right)$$

$$= \$200\left(e^{1.0} - 1\right)/\left(e^{0.1} - 1\right)$$

$$= \$3,268.$$

(c) What is the future worth including the compounding of interest?

$$(F/A_1, b, r, n) = \left\{[(e^{rn} - e^{bn})]/\left[\left(e^r - e^b\right)\right]\right\} \tag{9.61}$$

$$F = A_1(F/A_1, b, r, n)$$

$$= \$200\left(e^{10\times0.05} - e^{10\times0.10}\right)/\left(e^{0.05} - e^{0.10}\right)$$

$$= \$200(-1.069560)/(-0.05390)$$

$$= \$3,969.$$

(d) Therefore, the total interest earned over the 10 years would be:

$$F - TP = \$3,969 - 3,268 = \$701.$$

(e) What is the present worth?

$$(P/A_1, b, r, n) = A_1 \left\{ \left[1 - \left(e^{bn} / (e^{rn}) \right) \right] / \left[\left(e^r - e^b \right) \right] \right\}$$

(9.60)

$$P = A_1 \left\{ \left[1 - \left(e^{bn} / (e^{rn}) \right) \right] / \left[\left(e^r - e^b \right) \right] \right\}$$

$$= \$200[-0.64872/ - 0.05390]$$

$$= \$2{,}407.$$

(f) What is the equivalent annual uniform series payment A?

$$(A/A_1, b, r, n) = \left\{ \left[\left(e^{rn} - e^{bn} \right) \right] / \left[e^r - e^b \right] \right\} \{ (e^r - 1) / [e^{rn} - 1] \}$$

(9.62)

$$A = A_1 \left\{ \left[\left(e^{rn} - e^{bn} \right) \right] / \left[e^r - e^b \right] \right\} \{ (e^r - 1) / [e^{rn} - 1] \}$$

$$= 200\{ [-1.06956/ - .05390][0.05127/0.64872]$$

$$= 314.$$

9.9 ESCALATION GRADIENT CONTINUOUS COMPOUNDING FORMULAS

The escalation gradient starts with the gradient in the first period whereas the geometric gradient does not start the gradient until the second period. The factors for the discrete interest model will be converted to continuous interest factors. The factors will be developed for the case where $r \neq E$ and it will be left to the students to develop the case where $r = E$. The escalation gradient is used primarily in the construction industry and used with constant dollars. The symbol E is used for the escalation gradient and is similar to that of g in the geometric gradient. The factors for the discrete case were:

$$(P_E/A_1, E, i, n) = \left\{ [(1 + E)/(E - i)] \times \left[[(1 + E)/(1 + i)]^n - 1 \right] \right\}$$

(9.34)

$$(F_E/A_1, E, i, n) = \left\{ [(1 + E)/(E - i)] [(1 + E)^n - (1 + i)^n] \right\}$$

(9.38)

$$(A_E/A_1, E, i, n) = \left\{ [i(1 + E)/(E - i)] [(1 + E)^n - (1 + i)^n] / [(1 + i)^n - 1] \right\}.$$

(9.42)

Similarly, one would need relationships to convert the discrete gradient interest (E) to a continuous interest rate which will use the symbol (c). Therefore:

$$E = e^c - 1$$

(9.63)

and

$$(1 + E) = e^c$$

(9.64)

and thus

$$(1 + E)^n = e^{cn} \tag{9.65}$$

and

$$c = \ln(1 + E). \tag{9.66}$$

The conversion of these factors to continuous compounding of the interest and the gradient results in:

$$(P_E/A_1, c, r, n) = \{[(e^c)/(e^c - e^r)] \times [[(e^{cn})/(e^{rn})] - 1]\} \tag{9.67}$$
$$(F_E/A_1, c, r, n) = \{[(e^c)/(e^c - e^r)] \times [(e^{cn} - e^{rn})]\} \tag{9.68}$$
$$(A_E/A_1, c, r, n) = \{[(e^r - 1)(e^c)/(e^c - e^r)] \times [(e^{cn} - e^{rn})]/[(e^{rn} - 1)]\}. \tag{9.69}$$

9.9.1 ESCALATION GRADIENT CONTINUOUS INTEREST EXAMPLE

What would be the value at the end of 10 years of a escalation gradient of 10%($c = 10\%$) if the initial amount was $200 for a period of 10 years. The first payment would be at the end of the first year and the last payment at the end of the 10th year. The interest rate is 5%($r = 5\%$). (Note: if $c = 10\%$, $E = 10.517\%$.)

(a) What is the final payment? $200 $e^{0.10 \times 10}$ = $200(2.71828) = $544.

(b) What is the total payments (*TP*) of the gradients not including the interest? The expression is converted from the discrete form to the continuous form.

$$PT_E = A_1(1 + E)((1 + E)^n - 1)/E \tag{9.45}$$
$$= A_1(e^c)(e^{cn} - 1)/(e^c - 1)$$
$$= \$200(1.10517)(2.71828 - 1)/(1.10517 - 1)$$
$$PT_E = \$3,612.$$

(c) What is the future worth including the compounding of interest?

$$(F_E/A_1, c, r, n) = \{[(e^c)/(e^c - e^r)] \times [(e^{cn} - e^{rn})]\} \tag{9.68}$$
$$F_E = A_1\{[(e^c)/(e^c - e^r)] \times [(e^{cn} - e^{rn})]\}$$
$$= \$200[(1.10517/(1.10517 - 1.05127)] \times (2.71828 - 1.648720)$$
$$= \$200(1.10517/0.0539)(1.06956)$$
$$= \$4,386.$$

(d) Therefore, the total interest earned over the 10 years would be:

$$F_E - PT_E = \$4,386 - \$3,612 = \$774$$

(e) Determine the present worth

$$(P_E/A_1, c, r, n) = \{[(e^c)/(e^c - e^r)] \times [(e^{cn})/(e^{rn}) - 1]\} \qquad (9.67)$$
$$P_E = A_1 \{[(e^c)/(e^c - e^r)] \times [(e^{cn})/(e^{rn}) - 1]\}$$
$$= \$200[(1.10517/(1.10517 - 1.05127)] \times [.64872]$$
$$= \$200[13.301]$$
$$= \$2,660.$$

(f) What is the equivalent annual escalation payment A_e?

$$(A_E/A_1, c, r, n) = \{[(e^r - 1)(e^c)/(e^c - e^r)] \times [(e^{cn} - e^{rn})]/[(e^{rn}) - 1]\} \qquad (9.69)$$
$$A_E = A_1 \{[(e^r - 1)(e^c)/(e^c - e^r)] \times [(e^{cn} - e^{rn})]/[(e^{rn} - 1)]\}$$
$$= \$200[0.05127)(1.10517)/(1.10517 - 1.05127)]$$
$$\times [(2.71828 - 1.6487)/(0.6487)]$$
$$= \$200[1.05124] \times [1.06958/0.64872]$$
$$= \$347.$$

Note that the escalation gradient continuous compounding is larger than the escalation geometric gradient because of the additional gradient period. The continuous compounding formulas are presented in Table 9.2. Note that the values of the escalation geometric continuous compounding gradient are greater than the geometric continuous compounding by the factor when $c = b$ by $e^{c=b}$ which is the continuous compounding interest rate plus one—that is, A for the geometric gradient is 314 and $e^{0.1} = 1.10517$ and the A for escalation is $1.10517 \times 314 = 347$.

9.10 SUMMARY OF GRADIENT EXPRESSIONS

The discrete gradient expressions were derived in the first sections using discrete interest compounding and the results are summarized in Table 9.1. An example problem was solved in each section to illustrate the application of the formulas. The next four sections derived continuous interest compounding expressions and the results are summarized in Table 9.2. The difference in the results was relatively small and the continuous interest results tended to be larger for future worth and annual payments. The differences between the two compounding interest rates will be greater for a greater number of time periods and higher interest rates. This will happen in long range projects where life cycle costs are involved over periods such as 50–100 years in the life of a bridge, highway interchange, or large office building. The continuous interest is also

Table 9.2: Continuous compounding factors of economic expressions—discrete payments and continuous interest

Payment Type	Factor Name	Find	Given	Symbol	Formula
A. Single Payment					
	Present Worth	P	F	(P/F, r, n)	e^{-rn}
	Future Worth	F	P	(F/P, r, n)	e^{rn}
B. Uniform Payment					
(Uniform Series)	Sinking Fund	A	F	(A/F, r, n)	$[(e^r-1)/(e^{rn}-1)]$
	Capital Recovery	A	P	(A/P, r, n)	$[e^{rn}(e^r-1)/(e^{rn}-1)]$
	Future Worth	F	A	(F/A, r, n)	$[(e^{rn}-1)/(e^r-1)]$
	Present Worth	P	A	(P/A, r, n)	$[(e^{rn}-1)/(e^{rn}(e^r-1))]$
C. Uniform Gradient Expressions					
Standard Uniform Gradient	Uniform Gradient Present Worth	P	G	(P/G, r, n)	$\{[(e^{rn}-1) - n(e^r-1)]/[(e^r-1)^2\, e^{rn}]\}$
	Uniform Gradient Future Worth	F	G	(F/G, r, n)	$\{[(e^{rn}-1) - n(e^r-1)]/[(e^r-1)^2]\}$
	Uniform Gradient Uniform Series	A	G	(A/G, r, n)	$\{[(e^{rn}-1) - n(e^r-1)]/[(e^r-1)(e^{rn}-1)]\}$
Uniform Ramp Gradient	Uniform Ramp Gradient Present Worth	P_R	G	(P_R/G, r, n)	$\{[(e^{r(n+1)}-1) - (n+1)(e^r-1)]/[(e^r-1)^2 e^{rm}]\}$
	Uniform Ramp Gradient Future Worth	F_R	G	(F_R/G, r, n)	$\{[(e^{r(n+1)}-1) - (n+1)(e^r-1)]/[(e^r-1)^2]\}$
	Uniform Ramp Gradient Uniform Series	A_R	G	(A_R/G, r, n)	$\{[(e^{r(n+1)}-1) - (n+1)(e^r-1)]/[(e^r-1)^2(e^{rn-1})]\}$
D. Geometric Gradient Expressions					
Geometric Gradient	Geometric Gradient Present Worth	P	A_1,b	(P/A_1 b, r, n)	$\{[1-(e^{bn}/e^{rn})]/[e^r - e^b)]\}$
			If b=r	(P/A_1, b=r, n)	n/e^r
	Geometric Gradient Future Worth	F	A_1,b	(F/A_1, b, r, n)	$\{[e^{rn}-e^{bn}]/[e^r - e^b)]\}$
			If b=r	(F/A_1, b=r, n)	$n/e^{r(n-1)}$
	Geometric Gradient Uniform Series	A	A_1,b	(A/A_1, b, r, n)	$\{[e^{rn}-e^{bn}]/[e^r - e^b]\}\{[(e^r-1)/(e^{rn}-1)]\}$
			If b=r	(A/A_1, b=r, n)	$[n((e^{rm})/(e^{rn}-1)] * [(e^r-1)/(e^r)]$
Escalation Gradient	Escalation Gradient Present Worth	P_E	A_1,c	(P_E/A_1, c, r, n)	$\{[((e^c)/(e^c-e^r))] * [(e^{cn} - e^{rn})/e^{rn}]$
			If c=r	(P_E/A_1, c=r, n)	n
	Escalation Gradient Future Worth	F_E	A_1,c	(F_E/A_1, c, r, n)	$\{[((e^c)/(e^c-e^r))] * [(e^{cn} - e^{rn})]$
			If c=r	(F_E/A_1, c=r, n)	ne^{rn}
	Escalation Gradient Uniform Series	A_E	A_1,c	(A_E/A_1, c, r, n)	$\{[[(e^r-1)(e^c)(e^c-e^r)] * [(e^{cn}-e^{rm})/(e^{rn}-1)]\}$
			If c=r	(A_E/A_1, c=r, n)	$\{[n(e^r-1)e^{rn}/(e^{rn}-1)]\}$

Notation:
P=Present Worth;   i = effective discrete interest rate per period;   A=uniform end-of-period payments;   n = number of periods;
F=Future Worth;   g=Geometric Gradient Rate;   G=Uniform Gradient Amount;   E = Escalation Gradient Rate;
A_1 = Initial Geometric Gradient Amount and Initial Escalation Gradient Amount

used in the evaluation of interest rates on certificates of deposits in many institutions to show higher return rates. The formulas will be utilized in many of the problems in this chapter.

9.11 REFERENCES

[1] Creese, Robert C. and Adithan, M., *Strategic Cost Analysis for Project Managers and Engineers*, New Academic Science Limited, Tunbridge Wells, UK, pp. 25–58, 2012. 113, 142

[2] Park, Chan S., *Contemporary Engineering Economics*, 2nd ed., Addison-Wesley, Menlo Park, CA, p. 803, 1997. 113

[3] Newnan, Donald G., Eschenbach, Ted G., and Lavelle, Jerome P., *Engineering Economic Analysis*, 11th ed., Oxford University Press, New York, p. 655, 2012. 113

[4] Park, Chan S. and Sharp-Bette, Gunter P., *Advanced Engineering Economics*, John Wiley & Sons, Inc., New York, pp. 38–128, 1990. 113, 129, 142

9.12 EVALUATIVE QUESTIONS

1. If the time period is 5 years and the discrete interest rate is 10%, find the following values for the uniform gradient:

   (a) $(P/G, i = 10\%, n = 5)$

   (b) $(F/G, i = 10\%, n = 5)$

   (c) $(A/G, i = 10\%, n = 5)$

2. If the time period is 5 years and the discrete interest rate is 10% and the geometric gradient is 5%, find the following values:

   (a) $(P/A_1, g = 5\%, i = 10\%, n = 5)$

   (b) $(F/A_1, g = 5\%, i = 10\%, n = 5)$

   (c) $(A/A_1, g = 5\%, i = 10\%, n = 5)$

3. If the time period is 5 years and the interest rate is 10% and the escalation rate is 5%, find the following values:

   (a) $(P_E/A_1, \mathsf{E} = 5\%, i = 10\%, n = 5)$

   (b) $(F_E/A_1, \mathsf{E} = 5\%, i = 10\%, n = 5)$

   (c) $(A_E/A_1, \mathsf{E} = 5\%, i = 10\%, n = 5)$

4. The cost of materials for a project is \$1,000,000 per year for the next 5 years. The interest rate (return rate) is expected to be 15% and the escalation rate is predicted to be 5%. What is the expected present worth of the materials for the project (a) using discrete interest calculations with escalation and (b) using continuous interest and escalation calculations?

5. A contractor is building a high rise apartment which will take 3 years to build. The contractor predicts that the materials will be \$3,000,000 per year. Consider these as end-of-year expenses. He expects the materials to escalate by 8% per year. His expected return rate is 20%.

   (a) What is the present worth of this project to the contractor?

   (b) What is the Future Worth of the project to the contractor?

   (c) What is the annual end-of-year payment?

   (d) If the project was considered as a geometric gradient, what is the annual end-of-year payment?

   (e) The annual payment for the gradient was a beginning of period payment, what is the amount?

6. If the time period is 5 years and the nominal interest rate is 10%, find the following values:

   (a) $(P/F, r = 10\%, n = 5)$

   (b) $(F/P, r = 10\%, n = 5)$

   (c) If $F = \$300$, find the value of P

7. If the time period is 5 years and the discrete interest rate is 10%, find the following values:

   (a) $(P/F, i = 10\%, n = 5)$

   (b) $(F/P, i = 10\%, n = 5)$

   (c) If $P = \$300$, find the value of F

8. If the time period is 5 years and the nominal interest rate is 10%, find the following values:

   (a) $(P/A, r = 10\%, n = 5)$

   (b) $(P/F, r = 10\%, n = 5)$

   (c) $(A/P, r = 10\%, n = 5)$

   (d) $(A/F, r = 10\%, n = 5)$

   (e) $(F/P, r = 10\%, n = 5)$

   (f) $(F/A, r = 10\%, n = 5)$

   (g) If $A = \$200$, find the values of F and P

(h) If $P = \$200$, find the values of F and A

9. If the time period is 5 years and the discrete interest rate is 10%, find the following values:

(a) $(P/A, i = 10\%, n = 5)$

(b) $(P/F, i = 10\%, n = 5)$

(c) $(A/P, i = 10\%, n = 5)$

(d) $(A/F, i = 10\%, n = 5)$

(e) $(F/P, i = 10\%, n = 5)$

(f) $(F/A, i = 10\%, n = 5)$

(g) If $A = \$200$, find the values of F and P

(h) If $P = \$200$, find the values of F and A

10. If the time period is 5 years and the nominal interest rate is 10%, find the following uniform gradient factors:

(a) $(P/G, r = 10\%, n = 5)$

(b) $(F/G, r = 10\%, n = 5)$

(c) $(A/G, r = 10\%, n = 5)$

11. If the time period is 5 years and the nominal interest rate is 10%, find the following uniform ramp gradient values:

(a) $(P_R/G, r = 10\%, n = 5)$

(b) $(F_R/G, r = 10\%, n = 5)$

(c) $(A_R/G, r = 10\%, n = 5)$

12. If the time period is 5 years and the discrete interest rate is 10% and the geometric gradient g rate is 6%, find the following values:

(a) $(P/A_1, g = 6\%, i = 10\%, n = 5)$

(b) $(F/A_1, g = 6\%, i = 10\%, n = 5)$

(c) $(A/A_1, g = 6\%, i = 10\%, n = 5)$

(d) If $A_1 = \$200$, what are the values for P and F?

13. If the time period is 5 years and the nominal interest rate is 10% and the geometric gradient b rate is 6%, find the following values:

(a) $(P/A_1, b = 6\%, r = 10\%, n = 5)$

(b) $(F/A_1, b = 6\%, r = 10\%, n = 5)$

(c) $(A/A_1, b = 6\%, r = 10\%, n = 5)$

(d) If $A_1 = \$200$, what are the values for P, A, and F?

14. If the time period is 5 years and the nominal interest rate is 10% and the escalation gradient c rate is 6%, find the following values:

(a) $(P_E/A_1, c = 6\%, r = 10\%, n = 5)$

(b) $(F_E/A_1, c = 6\%, r = 10\%, n = 5)$

(c) $(A_E/A_1, c = 6\%, r = 10\%, n = 5)$

(d) If $A_1 = \$200$, what are the values for P_E, A_E, and F_E?

15. If the time period is 5 years and the nominal interest rate is 10% and the geometric gradient b rate is 10%, find the following values:

(a) $(P/A_1, b = 10\%, r = 10\%, n = 5)$

(b) $(F/A_1, b = 10\%, r = 10\%, n = 5)$

(c) If $A_1 = \$200$, what are the values for P, A, and F?

16. If the time period is 5 years and the nominal interest rate is 10% and the escalation gradient c rate is 10%, find the following values:

(a) $(P_E/A_1, c = 10\%, r = 10\%, n = 5)$

(b) $(F_E/A_1, c = 10\%, r = 10\%, n = 5)$

(c) $(A_E/A_1, c = 10\%, r = 10\%, n = 5)$

(d) If $A_1 = \$200$, what are the values for P_E, A_E, and F_E

17. Show the derivation of the equations for the geometric gradient continuous interest formula when the continuous interest rate r and the continuous gradient b are equal.

18. Show the derivation of the equations for the escalation gradient continuous interest formula when the continuous interest rate r and the continuous escalation gradient c are equal.

19. Construct a table in a spreadsheet and calculate the expressions for the Uniform Gradient and the Uniform Ramp Gradient factors of $(P/G, i, n)$, $(F/G, i, n)$, $(A/G, i, n)$, $(P_R/G, i, n)$, $(F_R/G, i, n)$, and $(A_R/G, i, n)$ for $i = 10\%$, and $n = 1$–60 and $n = 100$. Compare the values calculated with those in the various reference books [1–4], but it may be difficult as the only ones usually given are $(P/G, i, n)$ and $(A/G, i, n)$.

20. Construct a table in a spreadsheet and calculate the geometric gradient expressions for the factors of $(P/A_1, g, i, n)$, $(F/A_1, g, i, n)$, and $(A/A_1, g, i, n)$, for $i = 10\%$, $g = 5\%$, and $n = 1$–60 and $n = 100$. Compare the values calculated with those calculated for Problem 2.

21. Construct a table in a spreadsheet and calculate the escalation gradient expressions for the factors of $(P_E/A_1, \mathsf{E}, i, n)$, $(F_E/A_1, \mathsf{E}, i, n)$, and $(A_E/A_1, \mathsf{E}, i, n)$, for $i = 10\%$, $\mathsf{E} = 5\%$, and $n = 1$–60 and $n = 100$. Compare the values calculated with those calculated for Problem 3.

22. Construct a table in a spreadsheet and calculate the expressions for the Uniform Gradient and the Uniform Ramp Gradient factors with continuous compounding of $(P/G, r, n)$, $(F/G, r, n)$, $(A/G, r, n)$, $(P_R/G, r, n)$, $(F_R/G, r, n)$, and $(A_R/G, r, n)$ for $r = 10\%$, and $n = 1$–60 and $n = 100$. Compare the values calculated with those in Problem 8.

23. Construct a table in a spreadsheet and calculate the geometric gradient expressions for the factors of $(P/A_1, b, r, n)$, $(F/A_1, b, r, n)$, and $(A/A_1, b, r, n)$, for $r = 10\%$, $b = 6\%$, and $n = 1$–60 and $n = 100$. Compare the values calculated with those calculated for Problem 13.

24. Construct a table in a spreadsheet and calculate the escalation gradient expressions for the factors of $(P_E/A_1, c, r, n)$, $(F_E/A_1, c, r, n)$, and $(A_E/A_1, c, r, n)$, for $r = 10\%$, $c = 6\%$, and $n = 1$–60 and $n = 100$. Compare the values calculated with those calculated for Problem 14.

CHAPTER 10

Depreciation Terms, Methods, and Systems

10.1 INTRODUCTION

Depreciation is one of the most important deductions that businesses make to recover their initial investment in facilities and equipment used to produce their products and to lower their taxes. Every enterprise has some depreciation as facilities such as offices and computers are depreciable assets as well as items such as race horses, breeding cattle, hogs, goats or sheep, amusement parks, manufacturing facilities, and equipment, and almost anything you can think of to produce a product or service is depreciable. Depreciation is the item which causes major corporations to keep two sets of books, one for the government for paying taxes and another for the stockholders to show their earnings and profits. The primary sources of references used for depreciation in this work is Publication 946, *How to Depreciate Property*, for 2016 Returns [1]. Depreciation is extremely complicated, but the basics presented will meet most of the practical cases that are encountered in preparing estimates for projects and project evaluations.

A brief review of some of the terms from Chapter 1 will be repeated as they are critical in the determination of depreciation and its impact upon profits and cash flows. The basic terms are as follows.

1. Revenues = Income or money generated from Sales of Products and Services, Royalties, Financial Investments, Gambling/Lottery Winnings, etc.

2. Expenses = Income or money consumed in production of Products and Services such as labor, materials, equipment, computers, etc.

3. Depreciation = An annual income tax deduction that allows for the recovery of the cost or other basis of certain property over the time you use the property. It is an allowance for the wear and tear, deterioration or obsolescence of the property. IRS Publication 946 gives details on the calculations for depreciation [2].

4. Tax Rate = A percentage amount applied to the profits to determine the amount of taxes.

   The amount of taxes and profits are determined from these terms by the relationships previously presented in Chapter 1:

$$\text{Taxes} = (\text{Revenues} - \text{Expenses} - \text{Depreciation}) \times \text{Tax Rate} \qquad (10.1)$$

$$\text{Profits} = \text{Revenues} - \text{Expenses} - \text{Taxes.} \qquad (10.2)$$

10.1.1 CASH FLOWS

Cash Flows Before Taxes (CFBT) considers only net cash flows. This analysis is used when taxes cannot be considered and one is considering only the expenses. In general, one can consider:

$$\text{CFBT (Cash Flows Before Taxes)} = \text{Revenues} - \text{Expenses.} \qquad (10.3)$$

Cash Flows After Taxes (CFAT), considers taxes as an additional expense and is a more realistic of the actual cash flows. CFAT is typically used rather than CFBT as taxes can be a large item in the cash flows. Taxes are generally calculated based on a percentage of taxable income, and this is generally often represented as 40% in the U.S., but many of the major companies in the U.S. pay no income taxes. Some oil-rich countries in the middle east have zero income taxes whereas in some of the Northern European countries the total taxes reach 70%, but in these countries the people generally are satisfied in paying higher taxes as they often provide government pensions and/or healthcare. In Denmark, the tax on a new automobile was 180% and thus one paid 2.8 times the car price to purchase a new car.

The income tax rate in the U.S. was as high as 90% in the 1950's, but was reduced during the last half of the 20th century to under 40% for individuals and corporate taxes were recently reduced to 20%. Although new tax laws were passed in 2017 for 2018, the details have not been available from the Internal Revenue Service. Very high tax rates are often imposed during wars to fund the war effort. The 40% presented in most problems represents an estimate based upon the total of the federal, state and local taxes and although federal taxes are decreasing, it may cause an increase in the state and local taxes. The taxable income is based upon the net cash flows less the depreciation expenses. Thus, the amount allowed for depreciation must be determined to calculate the taxable income and taxes paid:

$$\text{CFAT (Cash Flows After Taxes)} = \text{Revenues} - \text{Expenses} - \text{Taxes.} \qquad (10.4)$$

10.2 DEPRECIATION TERMS AND DEFINITIONS

Depreciation is the systematic allocation of the cost of a capital asset over the recovery period (useful life). *IRS Publication 946* [1–11], which can be downloaded from the IRS website at http://www.irs.gov, is the best publication on depreciation for most individuals and is updated yearly for the new tax laws. Depreciation can be applied to nearly everything, but one notable exception is land. There are a wide variety of terms applied in depreciation as to depreciable property, depreciation life, and depreciation techniques. Many counties have different methods for depreciation, but the acceptable methods used in the U.S. system of depreciation will be emphasized.

10.2.1 DEPRECIATION CLASSES OF PROPERTY

The two major classes of depreciable property are tangible property and intangible property. Some examples of each are as follows.

A. Tangible Property—property you can see and touch [3]

  1. Personal Property—assets such as automobiles, houses, buildings, machines, computer equipment, furniture, etc.

  2. Real Property—land and buildings erected or agricultural produce growing on the land. The land is not depreciable, but buildings erected on the land are depreciable.

B. Intangible Property—property that has value, but cannot be seen or touched [3]

  1. Intangible Property—it includes items such as goodwill, computer software, copyrights, royalty, franchises, patents, trademarks, trade names and permits and licenses.

10.2.2 RECOVERY PERIOD AND DEPRECIATION LIFE

Recovery period is the life used for determining the depreciation life for recovery of the asset. However, there can be different permissible recovery periods permitted by the tax codes as most companies want rapid recovery periods, but some want long recovery periods as their income increases with increasing asset value and depreciation reduces the asset value.

Depreciable items must have a useful life of one year or more, be used in business or used to produce income, and they lose value via obsolescence, wear and tear, or natural causes, and are not inventory, stock in trade, or investment property. There are different types of asset life and the major types of asset life for consideration in determining the recovery period and the terms often considered are useful life, physical life economic life, and class life. The key term is the class life and is defined by the Internal Revenue Service (IRS) in the U.S. Class life is the "number of years that establishes the property class and recovery period for most types of property under the General Depreciation Schedule (GDS) and the Alternative Depreciation Schedule (ADS)" [4]. The various assets are assigned a specific asset class and that asset class will have a recovery period for that class life.

The recovery periods used in the Modified Cost Recovery System-General Depreciation System (MARCS-GDS) for property classes are presented in Table 10.1. The recovery periods have been set into nine classes to make classification easier. The property class is the recovery period in years, that is a three-year property class implies that the recovery period is three years.

The MACRS-ADS and MACRS-GDS have specific recovery periods and the recovery period for the MACRS-GDS is usually less than that of the MACRS-ADS. The GDS class life values are generally less than the asset class life period and is the accelerated depreciation life.

Table 10.1: **MACRS-GDS** property classes and types of property [4]

Property Class Life Recovery Period	Examples of Types of Property
3-year property	Trailer units for over-the-road use; special tools for—manufacture of motor vehicles, fabricated metal products, glass products, and rubber products.
5-year property	Automobiles, taxis, buses, light trucks, heavy duty trucks; information systems-computers and peripherals; construction; manufacture of apparel; cutting of timber; manufacture of chemicals and allied products; manufacture of electronic components, products, and systems.
7-year property	Equipment for the manufacture of cement, glass products, primary ferrous metals, foundry products, fabricated metal products, electrical and non-electrical machinery, motor vehicles, ship and boat building; office furniture and fixtures. (If no life is established, a product is classified as 7-year property)
10-year property	Petroleum refining equipment; equipment for the manufacture of grain and grain mill products; ship and boat building dry docks.
15-year property	Pipeline transportation, water transportation, telephone distribution equipment; electrical utility nuclear power plant; municipal wastewater treatment.
20-year property	Electric, gas, water and steam utility services; gas utility production plants; electric utility hydraulic production plants; gas utility distribution plants.
25-year property	Water utilities, municipal sewer.
Residential Rental Property 27.5-year	Residential Structures (depreciated over 27.5 years).
Nonresidential Property 39-year	Buildings (depreciated over 39 years).

One major exception is the recovery period for automobiles which is five years under the ADS and the class life recovery period is only three years. This occurred as loans for new automobiles increased from three years to five years and the government decided the class life should also increase. The ADS class life values are equal or greater than the asset class life period. The recovery periods are presented in Table 10.2 for several asset classes.

Table 10.2: Class life and recovery periods for selected asset classes [5]

Asset Class	Asset Description	Class Life (Years)	Recovery Period (Years) General GDS*	Alternative ADS**
0.11	Office furniture	10	7	12
0.12	Information systems, including computers	6	5	5
0.13	Data handling, except computers	6	5	6
0.22	Automobile, taxis	3	5	5
0.23	Buses	9	5	9
0.24	Heavy general purpose trucks	5	6	6
1.21	Cattle, breeding or dairy	7	5	7
1.223	Racehorse, above 2 yrs. age	none	3	12
10	Mining	10	7	10
13	Offshore drilling	7.5	5	7.5
13.3	Petroleum refining	16	10	16
15	Construction	6	5	6
24.4	Manufacture of wood products	10	7	10
30.1	Manufacture of rubber products	14	7	14
30.2	Manufacture of plastic products	11	7	11
32.1	Manufacture of glass products	14	7	14
33.2	Manufacture of primary nonferrous metals	14	7	14
33.3	Manufacture of foundry products	14	7	14
33.4	Manufacture of primary steel mill products	15	7	15
34.0	Manufacture of fabricated metal products	12	7	12
35.0	Manufacture of electrical machinery and other mechanical products	10	7	10
36.0	Manufacture of electronic components, products	6	5	6
36.1	Semiconductor manufacture equipment	5	5	5
37.1	Manufacture of motor vehicles	12	7	12
37.2	Manufacture of aerospace products	10	7	10
39.0	Manufacture of athletic, jewelry, and other goods	12	7	12
48.4	Satellite space segment property (satellites)	8	5	8
49.4	Central steam utility production and distribution	28	20	28
49.3	Liquefied natural gas plant	22	15	22
51.0	Municipal sewer	50	25	50
80.0	Theme and amusement parks	12.5	7	12.5
*GDS = General Depreciation System **ADS = Alternative Depreciation System				

10.2.3 DEPRECIATION CONVENTIONS

The possible depreciation conventions are full-year, mid-year, mid-quarter, and mid-month. The full-year convention was used previously, but now is no longer in use because all depreciable purchases made during the first year are not on the first day of the year. Thus, mid-year convention is the most used convention as it considers that purchases are made throughout the year. This is used when all purchases are spread out through the year and only one depreciation rate is used for the year.

Only mid-year convention will be considered in detail. Mid-year convention increases the number of years for depreciation calculations. Thus, a 5-year mid-year depreciation will have only half of the depreciation in the 1st and 6th years as it assumes the purchases are made throughout the year. If most purchases are made at the end of the year, then the mid-quarter or mid-month conventions would be required.

10.3   TRADITIONAL METHODS OF DEPRECIATION

There are numerous methods of depreciation and only a few of the more commonly used methods will be presented in detail. The straight line method is the traditional method which gave a uniform amount of depreciation over the life of the asset. The declining balance method gave a uniform percentage of depreciation over the investment life which gave larger amounts initially, but which has the problem of not going to zero. The MACRS system is a combination of both systems, giving the higher initial amounts of depreciation via the declining balance method initially and then switching to the straight line method to fully depreciate the item. The production-based system is not used in the U.S., but is used in other parts of the world and is based on the amount of use of the facility. There are special depreciation methods, and in the U.S., the method "Section 179" permits expensing the entire purchase in the first year, but there are several restrictions. Each of the methods will be present in more detail.

The depreciation amounts are used to update the book value of the asset. The book value is the initial invest minus the sum of all the depreciations up to the end of the year under consideration. It can be expressed in equation form by:

$$BV_k = B - \sum_{i=1}^{K} D_i, \tag{10.5}$$

where

B = initial investment amount
D_i = depreciation in years i
BV_k = book value at end of year k.

10.3.1 STRAIGHT LINE DEPRECIATION METHOD

The straight line method gives a constant amount of depreciation per year and this is why it was preferred as it is the easiest method. The expression for straight line depreciation is:

$$D_k = (B - SV)/N, \tag{10.6}$$

where

D_k = depreciation amount for all years $k(k = 1, N)$
B = investment (purchase cost + installation costs of asset)
SV = salvage value at end of life of asset at N years (usually taken to be zero)
N = depreciable life of asset, years
k = year of interest.

The salvage value is taken as zero when using the MACRS straight line schemes, and the actual net disposal value would be treated as a capital gain (or loss) when the asset is disposed. Also, the prediction of a salvage value several years in the future is difficult and by taking the value as zero, the salvage monies received could be taken as a capital gain. In some cases, the cost of removal would be greater than any salvage value and would be difficult to predict, but could be considered as a capital loss. One must not list only the purchase price, but also any installation costs such as connecting utilities, preparing foundations and other necessary items to make the purchase operable as the total cost is depreciable.

10.3.2 DECLINING BALANCE DEPRECIATION METHOD

The declining balance method is a constant percent of depreciation of the book value. It is a faster depreciation method in the initial years than the straight line method. However, the depreciation amounts decreases in the later years and become less than straight line and the total depreciation never reaches zero.

The expression for declining balance depreciation, for full year depreciation, is:

$$D_k = (B) \times R(1 - R)^{k-1}, \tag{10.7}$$

where

k = year of interest
D_k = depreciation for year $k(k = 1, N)$
B = investment (purchase cost + installation costs of asset)
R = depreciation rate = (usually 150% or 200%)/$(N \times 100)$
N = depreciable life of asset, years
SV = salvage value (the salvage value is not included in the calculations).

If the life is 10 years (N) and the depreciation rate of 200% is used, then $R = 200/(10 \times 100) = 0.2$, where R is a decimal less than 1.0. The rate of 200% implies the value initially would be twice that of straight line depreciation and 150% implies that the initial value would be 1.5 times that of straight line method. However, the amount for declining balance method decreases but never becomes zero. This method is not used directly today as a separate depreciation method, but it is used in combination with straight line depreciation be presented later as part of the MACRS. Using the mid-year convention, the depreciation rate would be 100% or $r = 0.10$ for the first year.

For mid-year convention, the declining balance is a more difficult expression:

$$D_1(\%) = R/(100 \times N \times 2) \tag{10.8}$$

$$D_k(\%) = R\left[1 - \sum_{i=1}^{k-1} D(\%)_i\right]. \tag{10.9}$$

The amount of depreciation would be

$$D_k(\$) = B_{k-1} \times D_k(\%). \tag{10.10}$$

10.3.3 DEPRECIATION EXAMPLE

What is the depreciation amounts using mid-year convention for an investment of $10,000 with an asset life of 5 years using straight line and double declining balance (200%)? The straight line depreciation would be 1/5 or 20%, that is $2,000. The 200% declining balance would be 2×20 or 40%. With the mid-year convention, the first year values and last year values will be half of that calculated. Formulas (10.6) and (10.7) were used to calculate the values with:

$B = 10,000$
$N = 5$ depreciable life of asset, years (however it will take 6 years to fully depreciate the item as only 1/2 of the amount calculated is allocated in the first and last years)

For example, in year 3 of the DDB, the amount of depreciation would be using Equations (10.8), (10.9), and (10.10):

$$D_1(\%) = 200/(100 \times 5 \times 2) = 0.20$$
$$D_3(\%) = 0.40(1 - 0.20 - 0.32) = 0.192$$
$$D_3(\$) = \$10,000 \times (0.192) = \$1,920.$$

Assume the asset is disposed in the last year. The calculation results are in Table 10.3.

SL (Straight Line)

$= 1/5 = 20\%$

Year 1 $= \frac{1}{2}(0.20 \times 10,000) = \$1,000$

Year 2 $= 0.20 \times 10,000 = \$2,000$

Year 3 $= 0.20 \times 10,000 = \$2,000$

Year 4 $= 0.20 \times 10,000 = \$2,000$

Year 5 $= 0.20 \times 10,000 = \$2,000$

Year 6 $= \frac{1}{2}(0.20 \times 10,000) = \$1,000$

DDB (Double Declining Balance)

$= 2/5 = 40\%$

Year 1 $= \frac{1}{2}(0.40 \times 10,000) = \$2,000$ (20.00%)

Year 2 $= 0.40 \times 8,000 = \$3,200$ (32.00%)

Year 3 $= 0.40 \times 4,800 = \$1,920$ (19.20%)

Year 4 $= 0.40 \times 1,920 = \$1,152$ (11.52%)

Year 5 $= 0.40 \times 1,152 = \$691.20$ (6.912%)

Year 6 $= 1/2(0.40 \times 691.20) = \$207.36$

The double-declining balance method is a percentage of the book value, so after the first year the amount is 40% of the book value of the previous year. In the first and last years, the amount is only ½ as a result of being mid-year convention. However, in the MACRS systems, the switch over in the last years is to straight line and the DDB method is not used.

Table 10.3: Straight line and declining balance depreciations and asset book values

Year	Life Remaining ($) at Year End	Straight Line		Declining Balance	
		Depreciation	Book Value ($)	Depreciation ($)	Book Value ($)
0	5	0	10,000	0.00	10,000.00
1	4½	1,000	9,000	2,000.00	8,000.00
2	3½	2,000	7,000	3,200.00	4,800.00
3	2½	2,000	5,000	1,920.00	2,880.00
4	1½	2,000	3,000	1,152.00	1,728.00
5	½	2,000	1,000	691.20	1,036.80
6	0	1,000	0	207.36	829.44
Total		10,000		9,170.56	

The initial depreciation is much more rapid with the declining balance in the early years (years 1 and 2 for this example), but it never reaches zero. The MACRS uses declining balance until the amount of depreciation is less than or equal to that of the straight line method for the remaining life. The straight line depreciation is based upon the remaining investment and the remaining life, and the straight line depreciation for year 4 would be the book value divided at the end of year 3 by the remaining life for year 3 which is 2.5 years (for years 4 and 5, and 1/2 year of year 6):

$$\text{SL Depreciation (year 4)} = \frac{\$2,880 \text{ (book value at the end of year 3)}}{2.5 \text{ (2.5 years remain after year 3)}} = \$1,152.$$

This is the same amount ($1,152) as the declining balance amount for year 4, so the switch to straight line would be made at the end of year 3 on a straight line basis. The original straight line depreciation is no longer used $2,000, but a straight line depreciation over the remaining 2.5 years which is $1,152. Since the straight line method is used, the depreciation for years 4 and 5 will also be $1,152 and year 6 would be for ½ year and thus $576.

The total depreciation would be $2,000 + 3,200 + 1,920 + 1,152 + 1,152 + 576 = 10,000$.

10.4 THE MACRS DEPRECIATION SYSTEMS

The MACRS are the depreciation systems used for depreciation. The MACRS systems follow the declining balance method with a switch-over to the straight line method. A more detailed explanation of the MACRS systems are in Publication 946. The salvage value in the MACRS system is taken as zero in the calculation of the depreciation amounts.

The two MACRS systems are:

1. GDS—General Depreciation System

2. ADS—Alternative Depreciation System

10.4.1 MACRS-GDS RECOVERY PERIODS AND PROPERTY CLASSES

The GDS is generally used unless ADS is required by law or if the user wants a slower depreciation. The GDS has the faster depreciation schedule as more depreciation occurs in the earlier years than in the ADS. The detailed Recovery Period and Property Classes for the GDS are listed in Table 10.1. The 200% and 150% values refer to the declining balance method percentage amount used before the switch to the straight line depreciation method. The nine property classes and recovery periods most frequent used [6] are:

(1) 3-yr property (200%)
(2) 5-yr property (200%)
(3) 7-yr property (200%—any property that does not have a class life specified is considered to have a 7-yr class life.)
(4) 10-yr property (200%)
(5) 15-yr property (150%)
(6) 20-yr property (150%)
(7) 25-yr property (150%)
(8) Residential Rental Property (27.5-yr, straight line)
(9) Non-Residential Real Property (39-yr, straight line)

The GDS does permit other systems and they are used in special instances when accelerated depreciation is not preferred. Accelerated depreciation leads to lower taxes paid, but also leads to lower profits or even losses which may not be desired. A list of all the possible ADS [6] system is as follows.

1. 200% declining balance for 3-, 5-, 7-, and 10-yr property

2. 150% declining balance over a GDS recovery period for all property used in farming businesses (except real property) and for all other property in the 15-, 20-, and 25-yr property classes

3. straight line for 3-, 5-, 7-, 10-, 15-, 20-, and 25-yr property as well as the residential rental property and non-residential real property

4. 150% declining balance over an ADS recovery period for property in the 3-, 5-, 7-, and 10-yr property classes.

10.4.2 MACRS-ADS RECOVERY PERIODS AND PROPERTY CLASSES

The ADS system almost always results in equal or longer recovery periods than the GDS [6]. For example, the personal property without a specified class life is 12 years in the ADS system compared to the 7 years in the GDS system. Some of the differences are the following.

1. See Table 10.2 Class Life Asset Depreciation Range (ADR) System.

2. Any personal property without a class life specified is 12 years.

3. Any real property without a class life specified is 40 years.

There are certain instances where the ADS must be used instead of the GDS. The ADS is required for the following.

1. Any tangible property predominantly used outside the U.S.

2. Any tax-exempt use property (churches, non-profit organizations, etc.).

3. Any property predominantly used in farming or agricultural business.

4. Any imported property covered by the executive order of the President of the U.S.

The primary systems of ADS used in practice are as follows.

1. 150% declining balance over the ADS recovery period.

2. Straight Line over the GDS Recovery Period (farming).

3. Straight Line of the ADS recovery Period.

10.4.3 MACRS-GDS MID-YEAR RECOVERY PERIODS

The MACRS is a declining balance method with a switchover to straight line [7]. This is the most commonly used system. The MACRS assumes a zero salvage value for all cases. Table 10.4 gives the depreciation percentages for the various mid-year recovery periods for the MARCS-GDS system. Note that there are only certain recovery periods; that is 3, 5, 7, 10, 15, and 20 years and that the totals of the columns are always 100.00.

Other conventions are the mid-quarter convention and mid-month convention and must be used for certain investments or other conditions. For example, if 40% or more of the purchases are in one quarter, the mid-quarter convention must be used. The mid-month is primarily for non-residential real property, (e.g., railroad grading or tunnel bore) and residential rental property. The mid-quarter and ADS tables are in the Publication 946. Only the mid-year convention will be presented in detail.

MACRS-GDS Depreciation Example

Consider the case of a $10,000 investment, zero salvage value, and determine the MACRS-GDS values for 200% GDS method for 5-yr property.

Using the mid-year convention it will take 6 years as the first and last years will receive only ½ of a depreciation value calculated for that year. Also, if an asset is sold in a year, only ½ of the depreciation is applied for that year.

$R = 200/(5 \times 100) = 0.40$ that is 40% depreciation per year (Double Declining Balance) and the straight line rate $R = 1/5 = 0.20$ or 20%. Note that the 200% declining rate is double that of the straight line method at the start. The straight line value changes according to the book value and life remaining values of the previous year. The declining balance depreciation is the 40% of the previous Book Value and is not used after the switch over point when it is less than or equal to the straight line depreciation. The results are in Table 10.5.

If one takes the Amount Used column and divides by 100, note that one obtains the exact same percentages in Table 10.4 for the Double Declining (200%) Balance column for a recovery period of 5 years. Thus, if one takes the values for the five year recovery in Table 10.5 and uses the percentages as decimals and multiplies by the investment of $10,000, one obtains the MACRS depreciation values in Table 10.5 which is presented in Table 10.6.

10.4.4 MACRS-ADS MID-YEAR RECOVERY PERIODS

There are several MACRS-ADS recovery period tables and they are much larger as they have many more periods and the full tables can be seen in Publication 946. A section of two tables will be presented to illustrate the similar construction of the tables using the same mid-year conventions. One table illustrate the straight line convention and the other will illustrate the 150% declining balance method. The values presented in these tables then can be used to compare with MACRS-GDS values in Table 10.4. Table 10.7 lists the percentages for the MACRS-ADS mid-year convention for straight line depreciation.

Table 10.4: Depreciation percentages for MACRS-GDS mid-year (half-year) recovery periods [7]

Year	Recovery Period					
	(Double Declining (200%) Balance)			(150% Declining Balance)		
	3	5	7	10	15	20
1	33.33	20.00	14.29	10.00	5.00	3.750
2	44.45	32.00	24.49	18.00	9.50	7.219
3	14.81	19.20	17.49	14.40	8.55	6.677
4	7.41	11.52	12.49	11.52	7.70	6.177
5		11.52	8.93	9.22	6.93	5.713
6		5.76	8.92	7.37	6.23	5.285
7			8.93	6.55	5.90	4.888
8			4.46	6.55	5.90	4.522
9				6.56	5.91	4.462
10				6.55	5.90	4.461
11				3.28	5.91	4.462
12					5.90	4.461
13					5.91	4.462
14					5.90	4.461
15					5.91	4.462
16					2.95	4.461
17						4.462
18						4.461
19						4.462
20						4.461
21						2.231
Totals	100.00	100.00	100.00	100.00	100.00	100.000

Table 10.8 lists the percentages for the MACRS-ADS mid-year convention for the 150% declining balance depreciation. Tables 10.7 and 10.8 give only a small portion of the values for these ADS values, but in comparison of the 3, 5, and 7 year value of the GDS values in Table 10.4, they are lower in the initial years. It is also interesting to note that the ADS schemes have many more recovery periods than the ADS , such as the 2.5, 3.5, 4, 6, and 7.5 in addition to the 3, 5, and 7 years and this would result in much more record keeping.

Table 10.5: MACRS depreciation calculations from declining balance and straight line

| Year | Life | Declining Balance Depreciation | Straight Line Depreciation | MACRS Depreciation | |
| | Remaining | Amount ($) | Amount ($)** | Amount Used ($) | |
(End of Year)	End of Year	200% of Straight Line Initial Amount	Use Remaining Life (SL)	(Largest of DDB and SL Methods)	End of Year Book Value
0	5	0	0	0	10,000
1*	4½	2,000	1,000	2,000	8,000
2	3½	3,200	1,778	3,200	4,800
3	2½	1,920***	1,920***	1,920	2,880
4	1½	****	1,152	1,152	1,728
5	½		1,152	1,152	652
6	0		652	652	0

*First year depreciation is for only 1/2 year in mid-year convention.

**Straight line is determined by the book value of the previous year divided by the remaining life at the end of the previous year.

*** Switch-over Point.

**** Remaining Declining Balance Depreciation items do not need to be calculated as they will be equal or less than the Straight Line Depreciation.

MACRS-ADS Depreciation Example

Consider the case of a $10,000 investment, zero salvage value, and determine the MACRS-ADS depreciation values for the straight line and 150% declining balance methods for 5-yr property. The percent rate is determined by:

$$R = 150/(5 \times 100) = 0.30 = 30\%.$$

The result is 30% depreciation rate per year for the 150% Declining Balance MACRS-ADS compared to the 40% depreciation in the MACRS-GDS. The faster depreciation increases cash flows and is advantageous for most companies to use the GDS depreciation. The Straight Line depreciation is even slower than the 150% Declining Balance Method, as shown in Table 10.9.

Table 10.6: MACRS-GDS depreciation calculations

1 Year	2 [Table 10.4] Recovery Period 5 Years Depreciation (%)	3 Column 2 Expressed as a Decimal Multiplied by $10,000	4 [Table 10.5] MACRS-GDS Depreciation Amount in $
0	-	-	0
1	20.00	2,000	2,000
2	32.00	3,200	3,200
3	19.20	1,920	1,920
4	11.52	1,152	1,152
5	11.52	1,152	1,152
6	5.76	576	576

Table 10.7: Depreciation percentages for MACRS-ADS straight line depreciation [8]

Recovery Period in Years-Depreciation Amounts in Percentages								
Year	2.5	3	3.5	4	5	6	6.5	7
1	20.00	16.67	14.29	12.50	10.00	8.33	7.68	7.14
2	40.00	33.33	28.57	25.00	20.00	16.67	15.39	14.29
3	40.00	33.33	28.57	25.00	20.00	16.67	15.38	14.29
4		16.67	28.57	25.00	20.00	16.67	15.39	14.28
5				12.50	20.00	16.66	15.38	14.29
6					10.00	16.67	15.39	14.28
7						8.33	15.38	14.29
8								7.14
*IRS Publication 946, p. 75								

Table 10.8: Depreciation percentages for MACRS-ADS 150% declining balance depreciation [9]

Recovery Period in Years-Depreciation Amounts in Percentages								
Year	2.5	3	3.5	4	5	6	6.5	7
1	30.0	25.0	21.43	18.75	15.00	12.50	11.54	10.71
2	42.0	37.5	33.67	30.47	25.50	21.88	20.41	19.53
3	28.0	25.0	22.45	20.31	17.85	16.41	15.70	15.03
4		12.5	22.45	20.31	16.66	14.06	13.09	12.25
5				10.16	16.66	14.06	13.09	12.25
6					8.33	14.06	13.09	12.25
7						7.33	13.08	12.25
8								6.03
*IRS Publication 946, p. 85								

Table 10.9: MACRS-ADS 150% declining balance and straight line depreciation calculations with a 5-yr recovery period

1	2	3	4	5
	150% Declining Balance	150% Declining Balance	Straight Line	Straight Line
Year	[Table 10.8] Depreciation (%)	(Column2/100)*$10,000 =Depreciation Amount ($)	[Table 10.7] Depreciation (%)	(Column4/100)* $10,000 =Depreciation Amount ($)
0	-	-	-	-
1	15.00	1,500	10.00	1,000
2	25.50	2,550	20.00	2,000
3	17.85	1,785	20.00	2,000
4	16.66	1,666	20.00	2,000
5	16.66	1,666	20.00	2,000
6	8.33	833	10.00	1,000

10.5 OTHER DEPRECIATION METHODS

There are several other depreciation systems, but two of the most interesting methods are the Section 179 and the Production-Based methods. The Section 179 method is part of the U.S. code and is used to expense rather than depreciate costs. The Production-Based method is based upon the amount of use of the item and is not approved in the U.S. tax code.

10.5.1 SECTION 179 DEPRECIATION

This Special Depreciation Method is unique to the U.S. and is intended to give small and medium-size companies a method of more rapid depreciation [10]. The limits that can be expensed change almost yearly with significant increases in the amounts. The maximum amount of depreciation that could be allowed for 2010 was $250,000 and it increased yearly to $500,000 in 2016. When the total depreciation amount exceeds $2,010,000, the maximum limit is reduced dollar for dollar. The total amount of depreciation must be less than $2,510,000 during 2016 to use any of Section 179 Depreciation. There are special limitations on passenger automobiles or Sport Utility Vehicles (SUV). Enterprise Zone Businesses can have an increase of $35,000 for the Section 179 limit and have a reduction in the reduced dollar for dollar limit.

The asset must be used 100% exclusively for business use. If less than 100%, only that percentage used for business can be used and one must use the GDS system to calculate yearly amounts.

Section 179 Depreciation Examples:

1. Farmer Jimmy bought a tractor for $600,000 during the year and that was the only purchase he has made. He could have a Section 179 Depreciation of $500,000 and the basis for the remaining depreciation of the tractor would then be $100,000.

2. Farmer Rosalyn bought a small tractor for $120,000 during the year and that was the only purchase she has made. She could have a Section 179 Depreciation of $120,000 and the tractor would be fully depreciated.

3. Donnie Great American Farms bought a tractor for $700,000 during the year and had a total depreciation for all their purchases of $2,410,000 in 2010. Thus, the maximum amount that could be used for Section 179 would be $500,000 − ($2,410,000 − $2,010,000) or $100,000. The remaining depreciation for the tractor would be $700,000 − $100,000 or $600,000.

4. Consultant Betty has purchased a new computer system for $8,000 and that is the sole depreciable item she purchased. Betty uses the computer for business 80% of the time and 20% for personal use. She can take 80% of the $8,000 or $6,400 as a business depreciation expense.

10.5.2 PRODUCTION-BASED DEPRECIATION

The production-based depreciation method is used where the extent of use of equipment depends on the production quantity or volume of production. The production quantity would be based on the number of units the equipment could produce before it is worn out. The volume of production may be the volume of material extracted from a resource before the resource is depleted of the material. This system is similar to that used in the U.S. for depletion of assets, such as oil, natural gas, and minerals equipment. However, the production-based depreciation system is not accepted by the IRS in the U.S. even though an example appears in Publication 946 [11].

Production-Based Depreciation Examples:

1. A truck has a capacity of hauling 500,000 tons during its lifetime and the initial cost of the truck is $50,000. During the 3rd year, the truck hauls 75,000 tons. The depreciation for the 3rd year would be:

$$\text{Depreciation} = (75,000/500,000) \times \$50,000 = \$7,500.$$

2. A gold mine has an estimated deposit of 500,000 troy ounces of gold and 1,000,000 tons of rock need to be mined to process the gold. The purchase cost of the mine was $3,000,000. A mine drilling machine was purchased for $400,000 to do all the mining as its expect life is also 1,000,000 tons. What would be the drilling machine depreciation if in the first year it mined 150,000 tons?

$$\text{Depreciation} = (150,000/1,000,000) \times \$400,000 = \$60,000$$

3. The expected life of a rolling mill roll unit is 600,000 tons of steel. The unit costs $1,400,000 and the amount rolled in year 2 was 80,000 tons. What would be the depreciation for the second year?

$$\text{Depreciation} = (80,000/600,000) \times 1,400,000 = \$186,667$$

10.6 SUMMARY

Cash flows and profits are the two major items for the financial success of an enterprise, and depreciation expenses have a major impact upon them. Depreciation is the recovery of expense in the past and cash flows increase as depreciation increases. Thus, accelerated depreciation accelerates cash flows into the enterprise. The IRS Publication 946 is a primary guide for depreciation. The declining balance and straight line methods have been combined and are the basis of the MACRS depreciation systems. The Section 179 depreciation system allows small and mid-size companies recover much of their equipment investments under $500,000 in the first year.

The MACRS depreciation systems and the Section 179 depreciation system represent the majority of depreciation systems used in the U.S. The advantage of the Section 179 depreciation is that you do not need to keep records to determine depreciation over the investment life as you can often fully depreciate the investment in the first year. This makes it very attractive for small and mid-size companies.

10.7 REFERENCES

[1] IRS Publication 946—*How to Depreciate Property* (for 2016 Returns), Department of the Treasury, Internal Revenue Service, (download at http://www.irs.gov), p. 114. 145, 146

[2] IRS Publication 946—*How to Depreciate Property* (for 2016 Returns), Department of the Treasury, Internal Revenue Service, (download at http://www.irs.gov), p. 3. 145

[3] IRS Publication 946—*How to Depreciate Property* (for 2016 Returns), Department of the Treasury, Internal Revenue Service, (download at http://www.irs.gov), pp. 110–111. 147

[4] IRS Publication 946—*How to Depreciate Property* (for 2016 Returns), Department of the Treasury, Internal Revenue Service, (download at http://www.irs.gov), pp. 30–31. 147, 148

[5] IRS Publication 946—*How to Depreciate Property* (for 2016 Returns), Department of the Treasury, Internal Revenue Service, (download at http://www.irs.gov), pp. 100–109. 149

[6] IRS Publication 946—*How to Depreciate Property* (for 2016 Returns), Department of the Treasury, Internal Revenue Service, (download at http://www.irs.gov), pp. 29–38. 154, 155

[7] IRS Publication 946—*How to Depreciate Property* (for 2016 Returns), Department of the Treasury, Internal Revenue Service, (download at http://www.irs.gov), p. 71. 156, 157

[8] IRS Publication 946—*How to Depreciate Property* (for 2016 Returns), Department of the Treasury, Internal Revenue Service, (download at http://www.irs.gov), p. 75. 159

[9] IRS Publication 946—*How to Depreciate Property* (for 2016 Returns), Department of the Treasury, Internal Revenue Service, (download at http://www.irs.gov), p. 85. 160

[10] IRS Publication 946—*How to Depreciate Property* (for 2016 Returns), Department of the Treasury, Internal Revenue Service, (download at http://www.irs.gov), pp. 15–23. 161

[11] IRS Publication 946—*How to Depreciate Property* (for 2016 Returns), Department of the Treasury, Internal Revenue Service, (download at http://www.irs.gov), p. 111. 146, 162

10.8   EVALUATIVE QUESTIONS

1. An asset was purchased for $80,000 and it took $20,000 to prepare the site and install the equipment. The asset has a recovery period of 7 years and MACRS-GDS depreciation was used.

   (a) What is the depreciation amount for the 1st year?

   (b) What is the book value at the end of the 3rd year?

   (c) What is the depreciation amount for the 5th year?

   (d) What is the book value after the 6th year?

2. An asset has value of $100,000 and a recovery period of 3 years. Use MACRS-GDS depreciation and determine the depreciation amount and book value over the life of the investment.

3. Your company has purchased a large new tractor trailer truck (heavy duty truck). It has a basic cost of $180,000 and with additional options costing $20,000, so the cost basis for depreciation purpose is $200,000. Its market value at the end of 5 years is estimated as $30,000 and will be depreciated under the GDS.

   (a) What is the cumulative depreciation through the end of the 3rd year?

   (b) What is the MACRS depreciation in the 4th year?

   (c) What is the book value at the end of the 2nd year?

   (d) What is the book value at the end of the 5th year?

4. Your company has purchased a large new tractor trailer truck (heavy-duty truck). It has a basic cost of $180,000 with additional options costing $20,000, so the cost basis for depreciation purpose is $200,000. Its market value at the end of 5 years is estimated as $30,000 and it will be depreciated under the ADS with straight line depreciation. (A heavy-duty truck has life of 6 years under ADS, but assume 5 years for this problem.)

   (a) What is the cumulative depreciation through the end of the 3rd year?

   (b) What is the MACRS-ADS depreciation in the 4th year?

   (c) What is the book value at the end of the 2nd year?

   (d) What is the book value at the end of the 5th year?

5. You are a private consultant and purchased a new computer valued at $3,000. You decide to use the Section 179 method for the computer you purchased as this is the only equipment purchase for the year.

   (a) What is the 1st year depreciation amount?

(b) What is the 2nd year depreciation amount?

6. You have started a consultancy company and purchased a new computer system valued at $3,000. You decided to use MACRS-GDS depreciation for the computer.

   (a) What is the 1st year depreciation amount?

   (b) What is the 2nd year depreciation amount?

7. A longwall mining machine was purchased for $500,000 and is expected to mine 5,000,000 tons of coal during its life. The machine mined 400,000 tons the 1st year and 700,000 tons the 2nd year.

   (a) What is the amount of depreciation using the production-based system?
   Year 1_____
   Year 2_____

   (b) If the MARCS-GDS system is used, what would be the depreciation amounts for the first two years?
   What is its recovery period?____
   If the GDS system is used, what would be the depreciation amounts for the first two years?
   Year 1_____
   Year 2_____

8. President Trump has decided that he wants to have a 4-year MACRS-GDS depreciation scheme. You, as his chief tax adviser, are to determine the depreciation rates in 24 hours, or be fired. What are the rates for each of the 5 years since mid-year convention is used?

9. Since the 25-year class life is new, the rates are not yet in Publication 946 (2016 Version). Therefore, for the 150% class, calculate what the rates should be for the MACRS-GDS system for a 25-year life.

10. An asset has value of $100,000 and a recovery period of 3 years. Use MACRS-ADS (150%) depreciation and determine the depreciation amount and book value over the life of the investment. Compare results with question 2.

CHAPTER 11

The Impact of Loans upon Cash Flows, Taxes, and Profits

11.1 INTRODUCTION

The use of loans is frequently necessary to purchase capital equipment including machinery, computers, facilities, materials, and other items necessary to produce products or services required for the enterprise. Loans are important as they impact cash flows, profits, and taxes. The loan is repaid in payments which contain two major components—the principal portion which is the portion used to repay the loan balance and the interest which is the fee for the use of the capital borrowed. The first section will be to analyze loans to determine the two components of principal and interest and then focus on how these items impact the cash flows before taxes, taxes, cash flows after taxes, and profits. The loan interest is a depreciable expense and the principal is not a depreciable expense, but it reduces the cash flows. Therefore, it is critical to know both the principal and the interest portions of a loan. Many general references [1–3] exist on general methods of loans, but only the Principal Present Value Approach will be presented in this chapter with permission of the American Society for Engineering Education [4].

11.2 THE PRESENT VALUE OF PRINCIPAL APPROACH FOR DETERMINING THE PRINCIPAL AND INTEREST COMPONENTS OF A LOAN

A new approach has been developed to determine the amounts of interest and principal of a loan. In the repayment of loans, where the loan payment is usually fixed for each of the payment periods. The individual principal and interest components change each period, with the principal payment increasing each period and the interest portion decreasing each period. However, the present worth of the principal is constant for each period and thus once determined, it permits relatively simple calculations for the two components of principal and interest for each period. The concept that the principal present value is constant gives a better understanding of how loan payments work. This approach is thus called the Present Value of Principal Approach or the Principal Present Worth Approach. To illustrate this process an example will be presented. The nomenclature used is expressed in Table 11.1.

Table 11.1: Nomenclature for loan principal and interest components [4]

Symbol	Variable Description	Formula for Variable or Type of Unit Used
LV or P	Loan value or initial principal	
i	Interest rate (%)	Used in decimal form
n	Loan life	Life in years
A	Loan payment	$A = LV[i^*(1+i)^n/\{(1+i)^n - 1\}] = LV^*(A/P,i,n)$
t	Time period of interest	$t = 1,2,...n$
PVP	Present value of principal per period	$PVP = [(A - i^*LV)/(1+i)]$
P(t)	Principal per period t	$P(t) = PVP^*(1+i)^t = PVP^*(F/P,i,t)$
I(t)	Interest per period	$I(t) = A - P(t)$
PVI(t)	Present value of interest per period	$PVI(t) = I(t)/(1+i)^t = I(t)^*(F/P,i,n)$
U(t)	Unpaid balance of loan after t periods	$U(t) = LV - \sum_{t=1}^{t} P(t) = LV - (A-i^*LV)((1+i)^t-1)/i)$
		$= LV - (A-i^*LV)(F/A,i,t)$
IT(t)	Total interest paid to period t	$IT(t) = \sum_{t=1}^{t} I(t) = t^*A - PT(t) = t^*A - (A-i^*LV)(F/A,i,t)$
PT(t)	Total principal paid to period t	$PT(t) = \sum_{t=1}^{t} P(t) = LV - U(t) = (A-i^*LV)(F/A,i,t)$

11.3 EXAMPLE PROBLEM OF LOAN PROBLEM USING PRESENT VALUE OF PRINCIPAL APPROACH

Let us consider determining the values of interest and principal payments on a Loan of $10,000 (LV) with an interest rate (*i*) of 5% with 10 (*n*) yearly end-of-year payments. The initial values are:

$$P = LV = \$10{,}000$$
$$i = 5\% = 0.05$$
$$n = 10.$$

The initial calculated values are the loan payment (*A*) and the present value of the principal (*PVP*):

$$A = LV \times (A/P, i, n) = LV \times i \times 1 + i)^n/\{(1 + i)^n - 1\}] \tag{11.1}$$
$$= \$10{,}000[0.05 \times 1.05)^{10}/\{(1.05)^{10} - 1\}] = \$1{,}295.05$$
$$PVP = [(A - i \times LV)/(1 + i)] = LV \times i/(1 + i)) \times 1/((1 + i)^n - 1)] \tag{11.2}$$
$$= [(1{,}295.05 - .05 \times 10{,}000)/(1 + .05)] = \$757.19.$$

All remaining calculations are based on these two calculated values—A and PVP—and the input values of LV, i, and n and the time of interest, t:

$$P(t) = PVP(F/P, i, t) = PVP(1 + i)^t \tag{11.3}$$

$$I(t) = A - P(t) \tag{11.4}$$

$$PVI(t) = I(t)(P/F, i, t) = I(t)/((1 + i)^t \tag{11.5}$$

$$U(t) = LV(F/P, i, t) - A(P/F, i, t) = LV(1 + i)^t - A[(1 + i)^t - 1]/i \tag{11.6}$$

$$PT(t) = \sum_{t=0}^{t=t} P(t) = (A - i \times LLV)(P/F, i, t) = (A - i \times LV) \times [((1 + i)^t - 1)/i] \tag{11.7}$$

$$IT(t) = \sum_{t=0}^{t=t} I(t) = t \times A - PT(t) = t \times A - (A - i \times LV)(P/F, i, t)$$

$$= t \times A - (A - i \times LV) \times \left[((1 + i)^t - 1)/i\right]. \tag{11.8}$$

Some interesting information about the totals of the loan components occur. (See Table 11.2.) The most important item is that the present value of the principal is a constant. Observe that the present value of the interest and the present value of the principal sum to the total initial value of the loan. The total sum of the principal parts of the loan is the expected total value of the initial loan value. These items are not as apparent in the traditional approaches to the evaluations of loans.

Calculate the present worth of the interest payments. Since they are not the same, each needs to be calculate individually. Thus,

$$PWI(\text{total}) = PWI(1) + PWI(2) + \cdots + PWI(10),$$

where

$$PWI(n) = I(1)/(1 + 0.05) + I(2)/(1 + 0.05)^2 + \cdots + I(10)/(1. + 0.05)^{10} = 2{,}428.$$

As a check on the PVP use, one can consider it to be an escalation gradient (E) of 5% the with and interest rate (i) of zero and $A_1 = 757.19$, thus, from Chapter 9:

$$F_E = A_1 \{[(1 + E)/(E - i)][(1 + E)^n - (1 + i)^n]\} \tag{9.37}$$

$$= \$757.19 \{[(1 + 0.05)/(0.05 - 0)][(1 + 0.05)^{10} - (1 + 0)^{10}]\}$$

$$= \$757.19\{13.20679\}$$

$$= \$10{,}000.$$

11.4 LOANS WITH CASH FLOWS, DEPRECIATION, PROFITS, AND TAXES

The basic relationships between the revenues, expenses, cash flows before taxes, cash flows after taxes, taxes, gross profits, net profits, depreciation, loan interest, and loan principal will be re-

Table 11.2: Calculations for example Problem 1 using present value of Principal Approach

Period	Interest per Period	Principal per Period	PV of Interest per Period	PV of Principal per Period	Unpaid Balance	Total Interest Paid	Total Principal Paid
t	I(t)	P(t)	PVI(t)	PVP	U(t)	IT(t)	PT(t)
0	0.00	0.00	0.00	0.00	10,000.00	0.00	0.00
1	500.00	795.05	476.19	757.19	9,204.95	500.00	795.05
2	460.25	834.80	417.46	757.19	8,370.16	960.25	1,629.84
3	418.51	876.54	361.52	757.19	7,493.62	1,378.76	2,506.38
4	374.68	920.36	308.25	757.19	6,573.25	1,753.44	3,426.75
5	328.66	966.38	257.52	757.19	5,606.87	2,082.10	4,393.13
6	280.34	1,014.70	209.20	757.19	4,592.17	2,362.44	5,407.83
7	229.61	1,065.44	163.18	757.19	3,526.73	2,592.05	6,473.27
8	176.34	1,118.71	119.35	757.19	2,408.02	2,768.39	7,591.98
9	120.40	1,174.64	77.61	757.19	1,233.38	2,888.79	8,766.62
10	61.67	1,233.38	37.86	757.19	0.00	2,950.46	10,000.00
Totals	2,950.46	10,000.00	2,428.14	7,571.86			

viewed. The nomenclature for the cash flows is in Table 11.3. The equations for the relationships will be presented and then an example problem will be used.

The basic expression for Cash Flows Before Taxes is:

$$CFBT = \text{Revenues} + \text{Loan Value} - \text{Expenses}$$
$$CFBT = R + LV - E. \tag{11.9}$$

The basic expression for Cash Flows with a Loan is:

$$CFAL = \text{Cash Flows Before Taxes} - \text{Loan Cash Flow}$$
$$CFAL = CFBT - LCF. \tag{11.10}$$

The basic expression for Taxable Income with a Loan is:

$$TI = \text{Cash Flows Before Taxes} - \text{Interest Paid} - \text{Depreciation}$$
$$TI = CFBT - I - D. \tag{11.11}$$

Table 11.3: Nomenclature for cash flow analysis with loans and depreciation [4]

Symbol	Variable Description	Formula for Variable or Type of Unit Used
INV	Investment that is depreciable	An expense usually at time zero
n_D	Project life (years)	
t	Study period year	t = 0,1,2....np
R(t)	Revenue	
E(t)	Expense	Expenses would include initial investment
CFBT(t)	Cash flows before taxes	CFBT(t) = R - E(includes INV)
LV	Loan amount	LV = P (usually at time zero)
A(t)	Loan payment	A = LV*(P/A, i, n_L)
i	Loan interest rate	
n_L	Loan life (years)	t_L = 1,2,....n_L
LI(t)	Loan interest amount	Loan interest for each period t
PVP	Present value of loan principal	PVP=(A - i*LV)/(1+i)
LCF(t)	Loan cash flow	LV(t=0) and A(t=1,2..n_L) values thru loan life
LP(t)	Loan principal amount	LP(t) = A(t) - LI(t)
CFAL(t)	Cash flows after loan	CFAL(t) = CFBT(t) - LCF(t)
DR(t)	Depreciation rate for year t	Depreciation rates for investment
D(t)	Depreciation amount for year t	D(t) = DR(t)*INV
TI(t)	Taxable income	TI(t) = CFBT(t) - I(t) - D(t)
TR	Tax rate	Usually specified and use decimal form
TP(t)	Taxes Paid	TP(t) = TI(t)* TR
CFAT(t)	Cash flows after taxes	CFAT(t) = CFAL(t) - TP(t)
NP(t)	Net profits	NP(t) = TI(t) - TP(t) = TI(t)(1.0 - TR)
CFAT(t)	Cash flows after taxes(a check)	CFAT(t) = NP(t) + D(t) - P(t)

The Basic Expressions for Taxes Paid are:

$$TP = \text{Taxable Income} \times \text{Tax Rate}$$
$$TP = TI \times TR \tag{11.12}$$
$$TP = (\text{Cash Flows Before Taxes} - \text{Interest} - \text{Depreciation}) \times \text{Tax Rate}$$
$$TP = (CFBT - I - D) \times TR. \tag{11.13}$$

The net profits can be determined by:

$$NP = \text{Taxable Income} - \text{Taxes}$$
$$NP = TI - TI \times TR = TI \times (1 - TR) \tag{11.14}$$

or

$$NP = \text{Cash Flows Before Taxes} - \text{Interest} - \text{Depreciation}) \times (1 - \text{Tax Rate})$$
$$NP = (CFBT - I - D) \times (1 - TR). \tag{11.15}$$

The cash flows after taxes can be determined by:

$$CFAT = \text{Cash Flows After Loan} - \text{Taxes Paid}$$
$$CFAT = CFAL - TP \tag{11.16}$$

or by

$$CFAT = \text{Net Profits} + \text{Depreciation} - \text{Loan Principal}$$
$$CFAT = NP + D - P(t). \tag{11.17}$$

11.5   EXAMPLE PROBLEMS OF LOANS WITH CASH FLOWS, DEPRECIATION, TAXES, AND PROFITS

Example Problem A. China Electronics-USA wants to purchase a new set of tooling with a total cost including installation of $50,000. This tooling would have a MACRS-GDS recovery period of 5 years to determine the depreciation percentages and the study period would be for 6 years. A loan for $20,000 would be needed and it is planned to pay the loan off in 4 years with an interest rate of 10% and the loan payments for each of the 4 years would be $6,309.42. The project is expected to generate a revenue of $30,000 per year and the expected annual costs are $14,000. The desired rate of return for the company is 15%. Although the totals for the *CFAT* and *NP* are the same with zero required return considerations, they are quite different as to when they occur. They are very different when the 15% return is considered. The present worth of the cash flows is only $3,710 whereas the net profits are $13,242 and the payback period is 4 years. The calculations are in Table 11.4.

　　Example Problem B. Use the same information in Problem A but use Section 179 Depreciation instead of the MACRS Depreciation. The difference between the two depreciation rates for problems A and B is significant when the required rates of return are considered. The faster depreciation (Section 179 in this problem) gives higher cash flows and much lower profits. Thus, the Section 179 depreciation or other accelerated depreciation would be reported to the government and straight line depreciation would be reported to the stockholders as slower depreciation gives higher profits in the early years. In the first year the Section 179 depreciation

Table 11.4: Example Problem A—Loan with cash flows, profits, taxes, and MACRS depreciation

Year End	Revenue	Expense	Cash Flows	Loan Cash Flow	Cash Flow After Loan	Loan Interest	Depr. Rate	Depr. Amount	Taxable Income	Taxes Paid	Cash Flow After Tax	Cum CFAT(t)	Net Profits
(t)	R(t)	E(t)	CFBT(t)	LCF(t)	CFAL(t)	LI(t)	DR(t)	D(t)	TI(t)	TP(t)	CFAT(t)	ΣCFAT(t)	NP(t)
0	0	50,000	-50,000	20,000	-30,000	0	0	0	0	0	-30,000	-30,000	0
1	30,000	14,000	16,000	-6,309	9,691	2,000	20.00	10,000	4,000	1,600	8,091	-21,909	2,400
2	30,000	14,000	16,000	-6,309	9,691	1,569	32.00	16,000	-1,569	-628	10,318	-11,591	-941
3	30,000	14,000	16,000	-6,309	9,691	1,095	19.20	9,600	5,305	2,122	7,569	-4,023	3,183
4	30,000	14,000	16,000	-6,309	9,691	574	11.52	5,760	9,666	3,867	5,824	1,801	5,800
5	30,000	14,000	16,000	0	16,000	0	11.52	5,760	10,240	4,096	11,904	13,705	6,144
6	30,000	14,000	16,000	0	16,000	0	5.76	2,880	13,120	5,248	10,752	24,457	7,872
Total	180,000	134,000	46,000	-5,238	40,762	5,238	100	50,000	40,762	16,305	24,457		24,457

Present Worth (CFAT-0%)   = 24,457
Present Worth (CFAT-15%) =   3,710
Present Worth (NP-15%)   = 13,242

gave a loss of $21,600 whereas the MARCS gave a profit of $2,400. The present worth of the cash flows increases to $7,298 whereas the present worth of the profits decreases to $7,860 and the payback period is reduced to 3 years. Straight line depreciation would give more profits and lower cash flows as its depreciation rates are slower than the MACRS method. The Section 179 results are in Table 11.5

11.6 SUMMARY

This new approach to loan evaluations does not contain terms to powers of $(t-1)$ or $(t+1)$ that has been required in the previous published approaches for loan calculations. This approach better explains how loans work in that the present worth of the principal is constant for each period t which has not been mentioned or emphasized previously in the literature.

The effects of loan interest and depreciation upon profits and cash flows is large when there is consideration of required return or the time value of money. The effects of different depreciation methods was illustrated was also illustrated when comparing MACRS vs. Section 179 Depreciation The present worth of Cash Flows After Taxes and Profits will be the same if there is no required return or time value of money consideration, but there are large differences when the required return is considered. The focus is on cash flows as accelerated depreciation give higher cash flows and lower taxes, but also lower profits. The profits will be reported to the stockholders using straight line depreciation as they will be greater, but an accelerated depreciation will be used to report to the government for taxes.

11.7 REFERENCES

[1] Creese, Robert C. and Adithan, M., *Strategic Cost Analysis for Project Managers and Engineers*, New Academic Science Limited, Tunbridge Wells, UK, pp. 74–82, 2012. 167

[2] Park, Chan S., *Contemporary Engineering Economics*, 2nd ed., Addison-Wesley, Menlo Park, CA, p. 803, 1997.

[3] Newnan, Donald G., Eschenbach, Ted G., and Lavelle, Jerome P., *Engineering Economic Analysis*, 11th ed., Oxford University Press, New York, p. 655, 2012. 167

[4] Creese, R. C., Present value analysis of traditional loans, paper presented at *ASEE Annual Conference and Exposition*, Atlanta, Georgia, pp. 23.981.1–23.981.10, June 2013. `https://peer.asee.org/22366` 167, 168, 171

(Materials published with permission of American society)

11.8 EVALUATIVE QUESTIONS

1. A loan for $40,000 is made for a period of 10 years with a 4% interest rate. Determine the loan payments for end-of-period payments, present value of the principal per period, the

Table 11.5: Example Problem B—Loan with cash flows, profits, taxes and Section 179 depreciation

Year End	Revenue	Expense	Cash Flows	Loan Cash Flow	Cash Flow After Loan	Loan Interest	Depr. Rate	Depr. Amount	Taxable Income	Taxes Paid	Cash Flow After Tax	Cum CFAT(t)	Net Profits
	R(t)	E(t)	CFBT(t)	LCF(t)	CFAL(t)	LI(t)	DR(t)	D(t)	TI(t)	TP(t)	CFAT(t)	ΣCFAT(t)	NP(t)
0	0	50,000	-50,000	20,000	-30,000	0	0	0	0	0	-30,000	-30,000	0
1	30,000	14,000	16,000	-6,309	9,691	2,000	100.00	50,000	-36,000	-14,400	24,091	-5,909	-21,600
2	30,000	14,000	16,000	-6,309	9,691	1,569	0	0	14,431	5,772	3,918	-1,991	8,659
3	30,000	14,000	16,000	-6,309	9,691	1,095	0	0	14,905	5,962	3,729	1,737	8,943
4	30,000	14,000	16,000	-6,309	9,691	574	0	0	15,426	6,171	3,520	5,257	9,256
5	30,000	14,000	16,000	0	16,000	0	0	0	16,000	6,400	9,600	14,857	9,600
6	30,000	14,000	16,000	0	16,000	0	0	0	16,000	6,400	9,600	24,457	9,600
Totals	180,000	134,000	46,000	-5,238	40,762	5,238	100	50,000	40,762	16,305	24,457	24,457	24,457

Present Worth (CFAT-0%) = 24,457
Present Worth (CFAT-15%) = 7,298 = -30,000 + 24,191/(1+.15)^1 + 3,918/(1+.15)^2 + 3,729/(1+.15)^3 + 3,520/(1+.15)^4 + 9,600/(1+.15)^5 + 9,600/(1+.15)^6
Present Worth (NP-15%) = 7,860 = 0 - 21,600/(1+.15)^1 + 3,918/(1+.15)^2 + 8,943/(1+.15)^3 + 9,256/(1+.15)^4 + 9,600/(1+.15)^5 + 9,600/(1+.15)^6

principal payments each period, the interest payments per period, the cumulative principal for each period, the cumulative interest per period, and the unpaid balance at the end of each period.

2. A loan is taken for a flat in the metropolis of Morgantown. The home is priced at $550,000 and the mortgage is for $400,000 at 6% APR for 30 years and the payments are made monthly.

   (a) What is the mortgage payment?

   (b) What is the interest on the 125th payment?

   (c) What is the principal on the 125th payment?

   (d) What is the total interest paid on the loan during the 30 years?

   (e) What is the remaining principal amount after the 125th payment is paid?

   (f) What is the total interest paid after the 125th payment is paid?

3. Company WV Consolidated has purchased a new 3D printing machine for $300,000 and with a loan $200,000 at 6% interest for 5 years. The annual income (savings) from this machine is expected to be $120,000 and the annual expenses are expected to be $30,000. MACRS-GDS depreciation is used, the 5-year class life is used and the tax rate is 40%.

   (a) What is the amount of depreciation for the 4th year?

   (b) What is the book value after the 4th year?

   (c) What would be the income taxes due for the 4th year assuming this is the only machine?

   (d) What would be the total cash flows for the 4th year?

   (e) What would be the net profits for the 4th year?

   (f) What are the *CFAT* (15%) at the end of the project?

   (g) What is the present worth of the profits at 15% return on the project?

4. Resolve the problem in Section 11.3 and Table 11.2 using an investment of $12,000 instead of $10,000 and calculate the annual payment, the interest per period, the principal per period, the PV of principal per period, the unpaid balance each year, and the cumulative total interest paid and the total principal paid at the end of each period using a 15% interest rate.

5. Resolve the Example Problem A (Table 11.4) of Section 11.5 using an investment of $60,000 instead of $50,000 and a loan for $30,000 instead of $20,000 and calculate all the values in Table 11.4 as well as the Present Worth of the Cash Flows and the Present Worth of the Profits using a Minimum Acceptable Rate of Return (MARR) of 15%.

6. Resolve problem Example Problem A in Table 11.4 using straight line depreciation with the mid-year convention and compare the results with those of Problem A and Problem B.

PART III

Methods for Project Evaluation and Risk Analysis

CHAPTER 12

Basic Project Evaluation Techniques

12.1 INTRODUCTION

There are several basic methods for evaluating projects and two projects will be presented for comparison by these basic traditional techniques. In some cases the results will be different, and the evaluations can be either on cash flows or net profits. The first seven techniques are the methods that have been commonly used for project evaluation [1–3] are presented in this chapter and Chapter 13. The commonly used basic techniques for project evaluation presented in this chapter are:

1. Payback Period

2. Discounted Payback Period

3. Present Worth (PW) Analysis

4. Future Worth (FW) Analysis

5. Average Annual Equivalent (Average Annual Cost)

6. Return on Original Investment (ROI)

7. Return on Average Investment (RAI)

These techniques will be illustrated with a sample problems utilizing the data in Tables 12.1 and 12.2. The techniques can be used on a cash flow before taxes or a cash flow after taxes basis or on a net profit basis and the results between the two projects will be compared. The analysis, however, should be made on an after tax basis (CFAT) whenever possible as these are the preferred results. Some investors may prefer the profit basis rather than cash flows, but cash flows are typically utilized. In addition, since companies use straight line depreciation instead of accelerated depreciation for the stockholders, the stockholders may want to examine the depreciation used for the government taxes for evaluating the actual profits rather than the reported "fake news" profits.

There are two projects under consideration by the Jen-Nat company—one project is to invest $50,000 in an additive manufacturing machine to cut tooling costs and lead times for production. The second project is to invest $50,000 for an improved computer security system to prevent hacking. Only one of the projects can be approved for implementation. The required return is 15% and loans are required for both projects.

The additive manufacturing project (referred to as Project A) will allow for the creation of tooling much faster and can be depreciated at the three-year class level. It will require improved computer skills by engineering staff to fully benefit the use of the machine and thus will have higher expenses, but the more rapid tooling will generate more revenue. The life of the project would be 5 years and a loan for $20,000 at 10% interest for a 3-year is required. MACRS depreciation would be for a 3-year property. The data for the project is presented in Table 12.1.

The improved security system to prevent cybersecurity attacks (referred to as Project B) would provide more security as well as improve computer services. The life of the project is for 6 years and MACRS depreciation would be for a 5-year property. A loan for $20,000 at 10% interest for 4 years can be obtained as the project life is 6 years. The data security system project is presented in Table 12.2.

The data in Tables 12.1 and 12.2 includes the loans, the loan interest, the taxable income, the taxes paid, the total cash flows after taxes, the net profits, the total net profits, the discounted cash flows after taxes, and the discounted net profits. The MARR used is 15% for discounting the cash flows and the net profits. The total net profits and the total cash flows after taxes are equal even though the values in the individual periods are quite different when no discount rate is used. This is a good check for your calculations. However, the total discounted net profits and the total discounted cash flows after taxes are not equal. This is because the amounts of cash flows and profits are frequently different in the same time period. The difference in amounts per period result in different total amounts. Thus, since the discounted profits and discounted cash flows are different, one must make a choice in selecting a criteria for the evaluation of a project. If there is a salvage value, it can be treated as a revenue in the last period when the equipment is sold. The various techniques for the evaluation of the projects will be examined using the data of the two example projects.

12.2   PAYBACK PERIOD

The payback period is the year when the cumulative cash flows becomes positive. This technique is used for small investments at lower levels of management where decisions are made quickly and when the payback period typically is less than 3 years. It may also have a total initial funding limit, such as $10,000 or $100,000 depending upon the size of the company initial funding limits. This not used when the funding is in the $1,000,000 range or higher as payback period is not the critical issue is large projects.

12.2.1   TRADITIONAL PAYBACK PERIOD

The payback period occurs in the year when the cumulative cash flows after taxes becomes positive. The limiting payback period may be as short as 1 year, and more frequently is considered

Table 12.1: Cash flows of additive manufacturing project (Project A)

End of Year (t)	Revenue Flow R(t)	Expanse Flow E(t)	Cash Flow Before Taxes CFBT(t)	Loan Cash Flow LCF(t)	Cash Flow After Loan CFAL(t)	Loan Interest Amount LI(t)	Depreciation Rate DR(t)	Depreciation Amount D(t)	Taxable Income TI(t)	Taxes Paid TP(t)	Cash Flow After Taxes CFAT(t)	Total Cash Flow ΣCFAT(t)	Net Profits NP(t)	Total Net Profits ΣBO(t)	Discounted Net Profits Profits	Discounted Cash Flow Flows-AT	Cumulative Discounted CFAT ΣCFAT(t)	Cumulative Discounted Net Profits ΣNP(t)
0	0	50,000	-50,000	20,000	-30,000		0	0	0		-30,000	-30,000	0	0	0	-30,000	-30,000	0
1	50,000	30,000	20,000	-8,042	11,958	2,000	33.33	16,665	1,335	534	11,424	-18,576	801	801	697	9,934	-20,066	697
2	50,000	30,000	20,000	-8,042	11,958	1,396	44.45	22,225	-3,621	-1,448	13,406	-5,170	-2,172	-1,371	-1,643	10,137	-9,929	-946
3	50,000	30,000	20,000	-8,042	11,958	731	14.81	7,405	11,864	4,746	7,212	2,042	7,118	5,747	4,680	4,742	-5,187	3,734
4	50,000	30,000	20,000	0	20,000	0	7.41	3,705	16,295	6,518	13,482	15,524	9,777	15,524	5,590	7,708	2,521	9,324
5	50,000	30,000	20,000	0	20,000	0	0.00	0	20,000	8,000	12,000	27,524	12,000	27,524	5,966	5,966	8,487	15,290
Totals	250,000	200,000	50,000	-4,127	45,873	4,127	100	50,000	45,873	18,349	27,524		27,524	27,524	15,290	8,487		

Table 12.2: Cash flows of computer security project (Project B)

End of Year (t)	Revenue Flow R(t)	Expanse Flow E(t)	Cash Flow Before Taxes CFBT(t)	Loan Cash Flow LCF(t)	Cash Flow After Loan CFAL(t)	Loan Interest Amount LI(t)	Depreciation Rate DR(t)	Depreciation Amount D(t)	Taxable Income TI(t)	Taxes Paid TP(t)	Cash Flow After Taxes CFAT(t)	Total Cash Flow ΣCFAT(t)	Net Profits NP(t)	Total Net Profits ΣBO(t)	Discounted Net Profits Profits	Discounted Cash Flow Flows-AT	Cumulative Discounted CFAT ΣCFAT(t)	Cumulative Discounted Net Profits ΣNP(t)
0	0	50,000	-50,000	25,000	-25,000		0	0	0	0	-25,000	-25,000	0	0	0	-25,000	-25,000	0
1	30,000	12,000	18,000	-7,887	10,113	2,500	20.00	10,000	5,500	2,200	7,913	-17,087	3,300	3,300	2,870	6,881	-18,119	2,870
2	30,000	12,000	18,000	-7,887	10,113	1,961	32.00	16,000	39	15	10,098	-6,989	23	3,323	18	7,635	-10,484	2,887
3	30,000	12,000	18,000	-7,887	10,113	1,369	19.20	9,600	7,031	2,812	7,301	312	4,219	7,542	2,774	4,800	-5,683	5,661
4	30,000	12,000	18,000	-7,887	10,113	717	11.52	5,760	11,523	4,609	5,504	5,816	6,914	14,456	3,953	3,147	-2,536	9,614
5	30,000	12,000	18,000	0	18,000		11.52	5,760	12,240	4,896	13,104	18,920	7,344	21,800	3,651	6,515	3,979	13,265
6	30,000	12,000	18,000	0	18,000		5.76	2,880	15,120	6,048	11,952	30,872	9,072	30,872	3,922	5,167	9,146	17,187
Totals	180,000	122,000	58,000	-6,547	51,453	6,547	100.00	50,000	51,453	20,581	30,872		30,872	30,872	17,187	9,146		

to be 2 or 3 years. Longer periods are generally not advised as the time value of money is usually not considered. Comparing the two alternatives from the data in Tables 12.1 and 12.2, the traditional payback periods are presented in Table 12.3.

Table 12.3: Traditional payback period

Payback Method	Project A (years)	Project B (years)
CFAT	3 years (2.72 Years)	3 years (2.96 years)
Project Life	5 years	6 years

The payback values as integers assume the year in which the first positive value occurs. The periods in parenthesis assume that cash flows are continuous throughout the year and, for Project A, would be calculated using the last negative and first positive values as:

$$\text{Project A Payback} = \text{year of last negative}$$
$$+ (-\text{CFAT(negative)}/(-(\text{CFAT negative}) + \text{CFAT Positive})) \quad (12.1)$$

$$= 2 + (-(5{,}170))/((-(-5{,}170)) + 2{,}042))$$
$$= 2 + (5{,}170)/(5{,}170 + 2{,}042)$$
$$= 2 + 0.72$$

Payback Period = 2.72 years.

Although the payback is frequently calculated in fractional years, the basis of cash flow analysis is the payments are at the end of the period and that payments are considered as discrete, not continuous. The payback period should be considered as three years in both cases and another criteria should be used. If only one alternative is being considered, then if payback period is less than or equal to the specified limit, it should be approved and if more than the specified limit, it should be rejected.

Some problems with the payback period analysis are as follows.

1. No consideration is given to the benefits after the payback period.

2. When comparing two investments, if the alternatives have different project life periods, the payback periods would be expected to be different. For this example, the projects would be repeated—6 times for Project A and 5 times for Project B to have the same life of 30 years.

3. The magnitude of the cumulative cash flows makes no difference, only the payback year—unless one is considering uniform continuous payments throughout the year.

Since Project A has a shorter payback period in uniform continuous payments throughout the year it would be the preferred project only if uniform continuous payments considerations are accepted. Otherwise, Project A and Project B have equal payback periods and another alternative project evaluation technique, such as return on investment, should be considered to differentiate between the projects if both cannot be approved as both meet the three year limit.

Note that discounted net profits would be a poor indicator of project performance as with the accelerated depreciation schemes, the 2nd year usually has much lower profits (and can be negative) than the 1st year which tends to be positive. This is another reason why cash flows are a better measure for project performance than profits.

12.2.2 DISCOUNTED PAYBACK PERIOD

A more realistic payback period can be determined by discounting the future cash flows and this is the basis of the Discounted Payback Method. Different paybacks consider the cumulative cash flows of the project until a positive cash flow results. The paybacks of interest are the cumulative discounted cash flows before taxes, the cumulative discounted cash flows after loan, and the cumulative the discounted cash flows after taxes. These discounted cash flows are presented in Table 12.4 using a discount rate of 15% for Project A.

Table 12.4: Project A payback periods by cumulative discounted cash flows for CFBT, CFAL, and CFAT

Year	Cash Flows Before Taxes	Discounted CFBT Flows	Cumulative Discounted CFBT Flows	Cash Flows After Loan	Discounted CFAL Flows	Cumulative Discounted CFAL Flows	Cash Flows After Taxes	Discounted CFAT Flows	Cumulative Discounted CFAT Flows
0	-50,000	-50,000	-50,000	-30,000	-30,000	-30,000	-30,000	-30,000	-30,000
1	20,000	17,391	-32,609	11,958	10,398	-19,602	11,424	9,934	-20,066
2	20,000	15,123	-17,486	11,958	9,042	-10,560	13,406,	10.137	-9.929
3	20,000	13,150	-4,336	11,058	7,862	-2,698	7,212	4,742	-5,187
4	20,000	11,435	+7,099	20,000	11,435	+8,737	13,482	7,708	+2,521
5	20,000	9,943	+17,042	20,000	9,944	+18,681	12,000	5,966	+8,487
Total	50,000	17,042		45,973	18,681		27,524	8,487	

Note that the discounted payback period of 4 years is the same for all 3 methods as all of them first become positive during the 4th year. The discounted payback period has the same problems as the traditional payback period. The discounted payback period typically increases the payback period when uniform continuous cash flows are assumed. When end-of-period cash flows are assumed, it may increase the payback period by one year or more over that of the traditional payback period. This is expected as the future cash flows are mainly positive and are being reduced by the discounting, so the payback period would tend to increase. The best evaluation technique for determining the payback period is the cumulative discounted cash flows after taxes.

12.3   TIME VALUE OF MONEY ANALYSIS FOR PROJECT PROFIT EVALUATION

The time value of money approaches probably the most utilized approaches for project analysis. The present worth, future worth, and average annual worth of the profits and cash flows, such as the CFAL, CFBT, and CFAT, are commonly used. The factor for discount is $(1 + i)^{-n}$ is to discount all of the n periods back to time zero to calculate the present worth which can then easily be converted to a future worth or average annual worth.

12.3.1   PRESENT WORTH ANALYSIS OF PROFITS

To determine the return on investment the returns use the profits rather than the cash flows. The present worth method discounts all the payments back to time zero. The present worth is the most commonly used method to determine the various cash flows or the profits. For profits and cash flows, one generally wants to maximize the value. The discount factor for discounting is $(1 + i)^{-n}$ where n is the number of periods the specific amount is to be discounted and i is the discounting factor. The present worth for the cash flows of the discounted net profits for Project A can be calculated using the data from Table 12.4 and the results are presented in Table 12.5.

The calculation procedure will be illustrated for calculating the present worth of the profits:

$$PW(NP - i\%) = NP(0) + (NP(1))/(1 + i/100)^1 + NP(2)/(1 + i/100)^2$$
$$+ NP(3)(1 + i/100)^3 + \cdots + NP(n)/(1 + i/100)^n \qquad (12.2)$$

$$PW(NP - 15\%) = 0 + 801/(1 + 0.15) - 2{,}172/(1 + 0.15)^2 + 7{,}118/(1 + 0.15)^3$$
$$+ 9{,}777/(1 + 0.15)^4 + 12{,}000/(1 + 0.15)^5$$
$$= 0 + 697 - 1{,}643 + 4{,}680 + 5{,}590 + 5{,}966$$
$$= +15{,}290$$

(if > 0.0 accept if a single alternative or if it is the greatest of all the alternatives considered).

The present worth approach has another major advantage in the it will give identical values when using actual currency or constant currency. One does not need to convert the dollars to constant dollars or use the inflation free interest rate for the base calculations. However, to obtain future worth or average annual equivalents, the constant dollars and inflation free interest rates must be used. The present worth analysis is the most commonly used method for project

Table 12.5: Project A discounted profits cash flows calculations data for determining present worth and ROI

Year	Net Profits NP	Discounted Net Profits (MARR=15%)	Cumulative Discounted Net Profits	Depreciation of Investment	Book Value (end-of-year) of Investment	Average Mid-Year Book Value of Investment During the Year**
0	0	0	0	0	50,000	50.000
1	801	697	697	16,665	33,335	41,667.5
2	-2,172	-1,643	-946	22,225	11,110	22,222.5
3	7,118	4,680	3,734	7,405	3,705	7,407.5
4	9,777	5,590	9,324	3,705	0	1,852.5
5	12.000	5,966	15,290	0	0	0
Totals	27,524	15,290		50,000	98,150	123,150

Include initial investment (50,000) which is used to calculate end-of-year book values

** Mid-Year Book Value = (Previous year book Value + Current year book Value)/2 and Initial Book Value

evaluations. Also note that the discount rate has a major impact upon the profits as shown by reducing the undiscounted profits from \$27,524 to the discounted profits of \$15,290.

12.3.2 FUTURE WORTH AND AVERAGE ANNUAL WORTH OF PROFITS

The results of the calculations for determining the present worth of the various cash flow expressions—CFBT, CFAL, and CFAT—are in Table 12.4. The present worth values can easily be converted to Future Worth Values or Average Annual Values, which is important for comparing projects which may have a different project evaluation periods, or commonly called the project life:

$$FW = F = P \times (F/P, i, n)$$
$$F = P \times (1 + i)^n$$
$$F = 15,290 \times (1 + 15/100)^5 = 15,290(1.15)^5$$
$$F = \$30,753, \tag{12.3}$$

where

$F = FW =$ future worth
$P = PW =$ present worth.

The future worth increases the values and those who are impressed by large numbers like to utilize this method. It is used to indicate future sums such as expected retirement incomes or pension values in the future.

The average annual value or average annual equivalent, A, is determined from:

$$AW = A = P \times (A/P, i, n)$$
$$= A = P((i/100) \times ((1 + i/100)^n)/((1 + i/100)^n - 1))$$
$$= A = 15{,}290 \times (0.15) \times ((1.15)^5/((1.15)^5 - 1))$$
$$= A = \$4{,}561, \tag{12.4}$$

where

$A = AW = $ Average Annual Worth $=$ Average Annual Equivalent
$P = PW = $ Present Worth.

The average annual equivalent values can be compared when the projects have different project life values and this is the best method for making decisions under those circumstances. If one uses the present worth analysis, one must have the same life and that results in multiple projects. For example, if Project 1 has a life of 4 years and the competing Project 2 has a life of 5 years, then a total life of 20 years would be needed. This results in repeating Project 1 five times and Project 2 four times to have an equivalent study period of 20 years. One of the evaluative questions requests the calculation of the average annual equivalent for Project B and compare it with Project A.

12.4   RETURN OF ORIGINAL INVESTMENT (ROI)

The return on original investment (often called return on investment—ROI) can be considered on a undiscounted (not discounted) or on a discounted basis. The return on investment is the average yearly profit divided by the initial fixed investment. This gives the average of the yearly ROI values. The data being applied is from Table 12.5

12.4.1 ROI – NOT DISCOUNTED

The basic form does not include discounting the profits. The formula used to obtain the values on a percentage basis is:

$$\text{ROI} = (\text{Average Yearly Profit/Original Fixed Investment}) \times 100. \tag{12.5}$$

For Project A the ROI would be:

$$\text{ROI} = \{[(801 - 2{,}172 + 7{,}118 + 9{,}777 + 12{,}000)/5)]/50{,}000\} \times 100$$
$$= \{[(27{,}524)/5]/50{,}000\} \times 100$$
$$= 11.01\%.$$

12.4.2 ROI – DISCOUNTED (ROI-D)

The total discounted profits were calculated previously as \$15,290 in Table 12.5 and thus the discounted ROI with the MARR of 15% would be:

$$\text{ROI-D} = (\text{Average Yearly Discounted Profit/Original Fixed Investment}) \times 100$$

$$= (\text{Total discounted Profits/5})/\text{Original Fixed Investment}) \times 100 \qquad (12.6)$$

$$\text{ROI-D} = \{[(15,290)/5]/50,000\} \times 100 = 6.12\%.$$

The ROI not-discounted percentage value is almost twice the percentage of the discounted ROI-D, and thus it is often preferred in the selling of the project. The annual worth of discounted profits with beginning-of-period payments is consistent in the timing of the payments and the investment. The discounted return is lower, but it is the return above the MARR value.

12.4.3 ROI ANNUAL WORTH – AW (ROI)

A new measure of the average annual worth of the profits would be with respect to the original investment when the investments lives are different or as another general approach to the ROI method. Since the annual worth is discounted, this represents the average return per year over the original investment.

In this case:

$$AW\ (ROI) = AW/I = 4,561/50,000 = 0.0912 = 9.12\%. \qquad (12.7)$$

12.4.4 ROI ANNUAL WORTH (BASE TIME) – AW-B (ROI)

Since the annual worth payments are end of periods and the investment occurs at time zero, the annual worth payments should be made beginning-of-period payments. This put both the annual worth payments and the original investment being considered at the same time (base time) for evaluating the return. Thus, for evaluating it at time zero, the annual worth payments must be moved to the beginning and is:

$$AW\text{-b}\ (ROI) = [AW/((1 + i))] \times I)$$

$$= [4,561/((1 + 0.15)] \times 50,000) = 0.0793 = 7.93\%. \qquad (12.8)$$

This rate [AW-b (ROI)] will be higher than the ROI-discounted as payments. All payments in the ROI discounted are taken at time zero whereas the AW-b (ROI) are considered yearly in the future at the beginning of the year.

12.5 RETURN ON AVERAGE INVESTMENT (RAI)

Since the original fixed investment is being depreciated over the investment period, the average investment over the investment life is thought to give a more reasonable representation of the investment amount. Both the actual investments and the discounted investments can be utilized in the calculations and each will be presented separately using the data in Table 12.5. The average investment is usually considered as the end-of-year book value, but it also can be considered as the average book value of the investment. Both calculations will be presented, with the end-of year book value first followed by the average book value.

12.5.1 RAI – NOT DISCOUNTED

The return on average investment (RAI) is the percentage relationship on the average annual profit to the average outstanding investment. The general formula used is:

$$RAI = (\text{Average Yearly Profit/Average Outstanding Fixed Investment}) \times 100, \qquad (12.9)$$

where

> RAI = Return on Average Investment using End-of-Year Book Values
> Average Yearly Profit = Total Net Profit/Total Project Life
> Average Outstanding Fixed Investment = Total End-of-Year Book
> Values/Total Project Life.

Therefore, using the values in Table 12.5:

$$RAI = (27{,}524/5)/(98{,}150/5) \times 100$$
$$= 28.04\%.$$

This is a rather large return compared to the ROI and is not representative of most operations. The sum of the end-of-year book values are less than the total investment, so an extremely large value is obtained. To obtain a better representation of the book value during the year, the mid-year book values are used which can be obtained by:

$$\text{Mid-Year Book Value} = (\text{Previous year book Value} + \text{Current year book Value})/2. \quad (12.10)$$

If the average mid-year book value is utilized, then the RAI values are somewhat more reasonable, but are still high. The revised form of Equation (12.7) is:

$$RAI\text{-}M = (\text{Average Yearly Profit/Average}$$
$$\text{Mid-Year Book Value of Fixed Investment}) \times 100, \qquad (12.11)$$

where

> RAI-M = Return on Average Investment using Mid-Year Book Value
> Average Yearly Profit = Total Net Profits/Total Project Life
> Average Mid-Year Book Value of Fixed Investment
>   = Total of Average Mid-Year Book Values of Investment/Total Project Life.

Therefore, using the values of Table 12.5:

$$RAI\text{-}M = (27,524/5)/(123,150/5) \times 100$$
$$= 22.35\%.$$

The 22.35% is still a high return on investment, but that is expected when the investment portion of the calculation is reduced. The next consideration will be to consider the reduction of the profits by discounting.

12.5.2 RAI – DISCOUNTED (RAI-D)

The discounted return on average investment (RAI-Discounted) is the percentage relationship on the average annual discounted profits to the average outstanding investment. The general formula used is:

$$RAI\text{-}D = (\text{Average Yearly Discounted Profit/Average}$$
$$\text{Outstanding Fixed Investment}) \times 100, \qquad (12.12)$$

where

    RAI-D = Return on Average Investment using Discounted Profits
    Average Yearly Discounted Profit = Total Net Discounted Profit/Total Project Life
    Average Outstanding Fixed Investment = Total End-of-Year Book
        Values/Total Project Life.

Therefore, using the values in Table 12.5:

$$RAI\text{-}D = (15,290/5)/(98,150/5) \times 100$$
$$= 15.58\%.$$

The discounted profits are the present worth of the profits and the RAI-D is much lower than the undiscounted RAI.

If the average book value is utilized, then the RAI-D values are somewhat more reasonable. The revised form of Equation (12.12) is:

$$RAI\text{-}DM = (\text{Average Yearly Discounted Profit/Average}$$
$$\text{Mid-Year Book Value of Fixed Investment}) \times 100, \qquad (12.13)$$

where

    RAI-DM = Return on Average Investment using Discounted Profits and
        Mid-Year Book Values
    Average Yearly Profit = Total Net Profits/Total Project Life

Average Mid-Year Book Value of Fixed Investment
= Total of Average Mid-Year Book Values of Investment/Total Project Life

Therefore, using the values of Table 12.5:

$$\text{RAI-DM} = (15{,}290/5)/(123{,}150/5) \times 100$$
$$= 12.42\%.$$

In the selection of a return on investment method, it is important to select one method and use it consistently. In selecting a method, it should be one that tends to best match your actual returns on investment. The study period should be over the economic life of the investment or when the book value first reaches zero or the expected salvage value. In general, the ROI values seem more reasonable as the book values decline more rapidly than the actual value of the equipment. Note that automobiles are fully depreciated in 5 years, but there are millions of cars on the road that are older than 5 years. However, the operating expenses of the automobile increase as they age.

12.6   SUMMARY

Several alternatives have been presented for evaluating projects. The best method presented thus far is probably the present worth method when projects have equivalent project lives, but the average annual worth is best when projects have different lives or when considering ROI analysis. The present worth method also has the advantage that it would result in the same results if used for constant dollars (inflation-free dollars) and inflation-free interest rates as well as actual dollars market rates including inflation.

The payback period method is based on the fact that earlier payback periods are usually better than later payback periods and would tend to reduce the risk of the project failing. This will also be shown by the project balance method illustrated in Chapter 13.

The average annual worth method is the best when comparing projects which have different project study periods. The present worth method would be used to calculate the values which would be converted to average annual worth values for comparison in the selection of the best method. The return on original investment is the used frequently as it does not depend on the discount rate of the returns or book values and reasonable values are obtained.

The rate of return methods provide alternative methods for project evaluation that can supplement the present worth overall approach. The present worth method has a pre-selected return for evaluation and does not determine the return automatically and all projects in companies do not have the same required return level. Projects critical to the survival of the enterprise may have lower return requirements than non-critical projects.

The ROI is a more used performance measure than the RAI methods, and the ROI-D better indicates if the required return is made on the original investment. The AW (ROI) method and beginning of period method, AW-b (ROI), appears best for evaluating ROI investments when the alternatives have different study periods. The AW-b (ROI) has both the AW payments and investment evaluated at the same time (beginning-of-period).

12.7 REFERENCES

[1] Creese, Robert C. and Adithan, M., *Strategic Cost Analysis for Project Managers and Engineers*, New Academic Science Limited, Tunbridge Wells, UK, pp. 83–108, 2012. 181

[2] Gelhausen, Marvin, Managing Editor, *Certification Study Guide*, 2nd ed., revised, AACE International, Inc, Morgantown, WV, pp. 17-1-6, 2003.

[3] Heinze, Kurt, *Cost Analysis of Capital Projects*, Marcel Dekker, Inc., New York, pp. 115–119, 1996. 181

[4] Humphreys, Kenneth K., *Jelen's Cost and Optimization Engineering*, McGraw-Hill, Inc., New York, pp. 103–110, 1991.

12.8 EVALUATIVE QUESTIONS

1. Prepare an equivalent Table 12.4 for Project B Payback Periods by Cumulative Discounted Cash Flows for CFBT, CFAL, and CFAT and determine the Project B Payback Periods by Cumulative Discounted Cash Flows for the CFBT, CFAL, and CFAT approaches. Are they different?

2. Determine the future worth value and average annual equivalents for Project B. Compare the values with Project A to make a selection.

3. Determine the end-of-year book values for Project A and Project B for each year over the life of the project.

4. Prepare a table similar to that of Table 12.5 for Project B. Then determine the discounted and non-discounted values of ROI and RAI; that is:

   (a) ROI

   (b) ROI-D

   (c) RAI

   (d) RAI-M

   (e) RAI-D

   (f) RAI-DM

   (g) AW (ROI)

   (h) AW-b (ROI)

Include initial investment (50,000) which is used to calculate end-of-year book values.

** Mid-Year Book Value = (Previous year book Value + Current year book Value)/2 and Initial Book Value.

5. Solve Project A using the MACRS-ADS straight line depreciation method (same number of years) and compare the results with the MACRS-GDS solution. Compare the payback periods the *PW* values of the profits and cash flows at 0% and 15% return.

6. Solve Project A using Section 179 depreciation method and compare the results with the MACRS solution. Compare the payback periods the *PW* values of the profits and cash flows in the 1st year and for the total project duration. Determine the various discounted and non-discounted values of ROI and RAI for the two methods; that is:

   (a) ROI
   (b) ROI-D
   (c) RAI
   (d) RAI-M
   (e) RAI-D
   (f) RAI-DM
   (g) AW (ROI)
   (h) AW-b (ROI)

7. An investment of $80,000 is made for a 3D-printing machine. It is expected to generate an annual revenue of $40,000 with annual expenses of $15,000. The project life is 8 years and the equipment has a class life of 5 years and MACRS depreciation is used. The 3D-printing machine will have a salvage value estimated at $5,000 when the project is complete at the end of the 8th year. The equipment has a class life of 5 years, MACRS depreciation will be used and the income taxes are 35% The required return is 10% and assume the capital gains tax is the same as the income tax. A loan for $20,000 is needed and the interest rate is 15%.

   Answer the following questions using the data from Table 12.6.

   (a) (i) Determine the Payback Period in years.
       (ii) Determine the Discounted Payback Period in years.
   (b) Determine the Present Worth of the project CFAT.
   (c) Determine the Average Annual Worth of the Project CFAT.
   (d) Determine the Returns on Investment—ROI and ROI-D, AW (ROI), and AW-b (ROI).
   (e) Determine the Returns on Average Investment—RAI, RAI-D, RAI-M, and RAI-DM.

Table 12.6: Data for Problem 7

Year	Revenue R	Investment I	Expenses E	CFBT	Loan Cash Flow	After Loan CFAL	Loan Interest	Depreciation %	Depreciation Amount	Taxable Income	Taxes Paid	Net Profits	CFAT	Cum CFAT	Disc CFAT	Cum DiscFAT
0	0	80,000	80,000	-80,000	20,000	-60,000	0	0	0	0	0	0	-60,000	-60,000	(60,000)	(60,000)
1	40,000		15,000	25,000	-8,760	16,240	3,000	20	16,000	6,000	2,100	3,900	14,140	-45,860	12,855	(47,145)
2	40,000		15,000	25,000	-8,760	16,240	2,136	32	25,600	-2,736	-958	-1,778	17,198	-28,661	14,213	(32,932)
3	40,000		15,000	25,000	-8,760	16,240	1,143	19.2	15,360	8,497	2,974	5,523	13,266	-15,395	9,967	(22,965)
4	40,000		15,000	25,000		25,000		11.52	9,216	15,784	5,524	10,260	19,476	4,080	13,302	(9,662)
5	40,000		15,000	25,000		25,000		11.52	9,216	15,784	5,524	10,260	19,476	23,556	12,093	2,430
6	40,000		15,000	25,000		25,000		5.76	4,608	20,392	7,137	13,255	17,863	41,419	10,083	12,513
7	40,000		15,000	25,000		25,000			0	25,000	8,750	16,250	16,250	57,669	8,339	20,852
8	40,000		15,000	25,000		25,000		0	0	25,000	8,750	16,250	16,250	73,919	7,581	28,433
9	5,000		0	5,000		5,000		0	0	5,000	1,750	3,250	3,250	77,169	1,516	29,949
Total	325,000	80,000	200,000	125,000	(6,279)	118,721	6,279	100	80,000	118,721	41,552	77,169	77,169		29,949	

CHAPTER 13

Advanced Project Evaluation Techniques

13.1 INTRODUCTION

The project evaluation techniques are mainly used to evaluate single projects on an accept-reject basis and are difficult for selecting the best of several of several projects when investing funds are limited. The techniques are the internal rate of return, the modified internal rate of return, benefit/cost ratio analysis, and project balance. These comparisons can be added to your spreadsheets in evaluating projects. General references [1–3] for this section have guided the arrangement of these topics.

The Internal Rate of Return (IRR) method determines the actual rate of return of the project and one can select the project with the highest rate of return. It was difficult to calculate the IRR before computers as it required several trial calculations, but with computers repeated calculations are performed very rapidly and this technique now more frequently applied.

The Modified Internal Rate of Return (MIRR) method, also referred to as the External Rate of Return (ERR) method, was used to closely approximate the IRR and was much easier to calculate as it could be performed in a single calculation. It utilizes the future worth of the benefits divided by the present worth of the costs to determine the MIRR, but the ease in now calculating the IRR has reduced the importance of the MIRR.

The Benefit/Cost (B/C) ratio method involves the similar calculations as the present worth method, but often involves other factors such as safety issues, environmental issues, and economic development factors which often are not considered in traditional present worth analysis. It is, however, another view of the present worth method by rearranging its components to measure the ratio of the benefits to the costs

The Project Balance (PB) method considers the value of the cash flow by escalating the initial cash flow (which usually are negative) through the period and adding the cash flow earned during the period at the end of the period. The final project balance and the end of the study period should be positive for the project to be acceptable and the area of the negative balances compared to the area of the positive balances give an indication of the risk of the project. Projects with no positive areas or small positive areas would be considered as risky projects. This is a more conservative approach than the other evaluation methods.

13.2   INTERNAL RATE OF RETURN (IRR)

The IRR is compared to the firm's MARR and if the IRR is greater than the MARR, the project is acceptable. One advantage is that the IRR can be compared to projects that have a different project life and/or a different investment amount. The IRR method calculates the specific rate of return that makes the present worth of the project cash flows equal to zero. Typically, the present worth of the cash flows is positive at a zero rate of return and is reduced as the rate of return increases and when it becomes zero, then the project would earn the desired rate of return.

The example Project A from Chapter 12 will be used as an example for calculations and the data is repeated in Table 13.1. The additive manufacturing project will allow the creation of tooling much faster and can be depreciated at the three year level. The investment is $50,000 and the required return is 15%. It will require improved computer skills by engineering staff to fully benefit the use of the machine and thus will have higher expenses, but the more rapid tooling will generate more revenue. The life of the project would be 5 years and a loan for $20,000 at 10% interest is needed to assist in the initial financing.

The present worth of the of the project cash flows with the 15% required return is $8,487. The return rate will be increased until the present worth of the project cash flows becomes zero and values are listed in Table 13.2. These calculations are now performed easily with spreadsheets and the IRR is 26.4% and are illustrated in Figure 13.1 This is the anticipated return that the project will bring. Thus, by comparing the IRR's, the project with the highest IRR would be preferred and the effects of project life and investment are considered. Discounting of net profit cash flows does not work for IRR evaluations as it would be positive even at 100% IRR. This is another reason why cash flows are preferred over profits for project evaluation analysis.

The primary disadvantage is that trial and error is typically used, but this is no longer a serious problem as it can be done automatically by the computer. There may be cases where there is more than one IRR, but the first one, which would be the lowest, is usually the one desired. The MIRR has only one value, so it was used as a check as an estimate of the IRR, but it is now rarely used as the actual IRR can be calculated directly with little difficulty.

13.3   MODIFIED INTERNAL RATE OF RETURN (MIRR)

The MIRR was used to estimate the IRR and has only one rate of return without the possibility of multiple rates of return. However, it is only an approximate value of the IRR but will be close and if multiple values of the IRR occur, the one nearest to the MIRR would be the correct one to select.

The MIRR method takes the future worth of the of the net positive cash flows and the present worth of the net negative cash flows to determine the MIRR and can be expressed as:

$$(1 + MIRR)^n = \frac{\text{Future Worth of positive cash flows at MARR}}{\text{Present Worth of negative cash flows at MARR}}. \tag{13.1}$$

Table 13.1: Cash flows of additive manufacturing project (Project A)

End of Year	Revenue Flow	Expanse Flow	Cash Flow Before Taxes	Loan Cash Flow	Cash Flow After Loan	Loan Interest Amount	Depreciation		Taxable Income	Taxes Paid	Cash Flow After Taxes	Total Cash Flow	Net Profits	Total Net Profits	Discounted		Cumulative Discounted	
							Rate	Amount							Net Profits	Cash Flow	CFAT	Net Profits
(t)	R(t)	E(t)	CFBT(t)	LCF(t)	CFAL(t)	LI(t)	DR(t)	D(t)	TI(t)	TP(t)	CFAT(t)	ΣCFAT(t)	NP(t)	ΣBO(t)	Profits	Flows-AT	ΣCFAT(t)	ΣNP(t)
0	0	50,000	-50,000	20,000	-30,000	0	0	0	0	0	-30,000	-30,000	0	0	0	-30,000	-30,000	0
1	50,000	30,000	20,000	-8,042	11,958	2,000	33.33	16,665	1,335	534	11,424	-18,576	801	801	697	9,934	-20,066	697
2	50,000	30,000	20,000	-8,042	11,958	1,396	44.45	22,225	-3,621	-1,448	13,406	-5,170	-2,172	-1,371	-1,643	10,137	-9,929	-946
3	50,000	30,000	20,000	-8,042	11,958	731	14.81	7,405	11,864	4,746	7,212	2,042	7,118	5,747	4,680	4,742	-5,187	3,734
4	50,000	30,000	20,000	0	20,000	0	7.41	3,705	16,295	6,518	13,482	15,524	9,777	15,524	5,590	7,708	2,521	9,324
5	50,000	30,000	20,000	0	20,000	0	0.00	0	20,000	8,000	12,000	27,524	12,000	27,524	5,966	5,966	8,487	15,290
Totals	250,000	200,000	50,000	-4,127	45,873	4,127	100	50,000	45,873	18,349	27,524		27,524		15,290	8,487		

Table 13.2: Internal rate of return calculations

Estimated IRR (%)	Cash Flows After Taxes
0	27,524
15	8,487
20	4,237
25	866
26	244
26.4	1

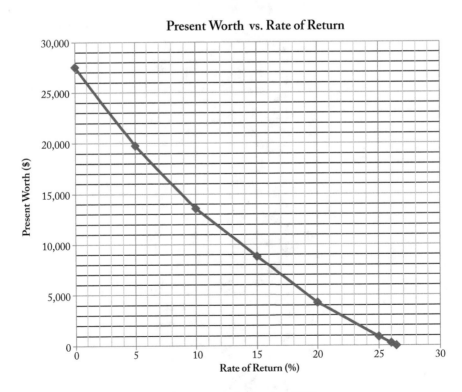

Figure 13.1: Present worth vs. rate of return to determine IRR.

Table 13.3: Positive future worth and negative present worth for calculating MIRR

Year	Positive Cash Flows	Future Worth of Positive Cash Flows	Negative Cash Flows	Present Worth of Negative Cash Flows
0			30,000	30,000
1	11,424	19,981		
2	13,406	20,389		
3	7,212	9,538		
4	13,482	15,504		
5	12,000	12,000		
Totals	57,524	77,412	30,000	30,000

Using the data of Table 13.3 in Equation (13.1) for the MIRR one obtains:

$$(1 + MIRR)^5$$
$$= \frac{11,424(1 + 0.15)^4 + 13,406(1 + 0.15)^3 + 7,212(1 + 0.15)^2 + 13,482(1 + 0.15) + 12,000}{30,000}$$
$$(1 + MIRR)^5 = 2.58039$$
$$(1 + MIRR) = 1.2089$$
$$MIRR = 20.9\%.$$

Note that the *MIRR* (20.9%) is greater than the MARR (15%), but not near to the actual IRR (26.4%) for this case. This is because the MIRR used the MARR to calculate the values and then it should be recalculated using the MIRR value and calculate a new MIRR. The MIRR is now rarely used as it is relatively easy to directly calculate the IRR.

13.4 BENEFIT/COST RATIO

B/C analysis utilizes the present worth procedure to determine the benefits and costs. There are two versions of the analysis—the conventional B/C ratio and the modified benefit cost M (B/C) ratio. The conventional ratio utilizes the total benefits and total costs as the ratio components whereas the M (B/C) utilizes the net benefits and total investment costs as the ratio. The M (B/C) will give higher values of the ratio, but the decision as to whether the process is acceptable will be the same.

13.4.1 CONVENTIONAL BENEFIT/COST RATIO

The B/C ratio is an extension of the present worth method as it uses the present worth value of the benefits compared to the present worth value of the costs. When the present worth of the benefits exceeds the present worth of the costs, the ratio of benefits to costs exceeds unity (1.0), and the project is considered acceptable. This is the same as the total present worth being greater than zero which is the criteria for an acceptable project by present worth analysis. This approach is frequently used for evaluation of government projects and other projects where the benefits and costs include the potential effects upon society, safety, and the environment, which are difficult to measure directly as well as the direct benefits and costs. The saving of lives, fewer accidents, etc. are considered as benefits and the costs for implementing them are considered as costs. The problem also requires the value of saving a life and administrations who prefer to avoid safety and environmental projects use a low value of a "life" and administrations who want to implement safety and environmental measures use a higher value of "life." The ratio is also dependent upon the rate of return and, in general, the higher the required rate of return, the less likely for the approval of the project as most benefits occur in the future, and higher MARR values decrease the present worth of the future benefits.

The acceptance of a project by present worth analysis requires that the project value is positive, that is:

$$PW \text{ (project value)} \geq 0. \tag{13.2}$$

For B/C analysis, the project is separated into benefits (positive values) and costs (negative values), and this results in Equation (13.2) being rewritten as:

$$PW \text{ (benefits)} - PW \text{ (costs)} \geq 0.$$

That is the approach of present worth analysis and the approach of B/C analysis is:

$$PW \text{ (benefits)} \geq PW \text{ (costs)}.$$

Now dividing PW (Benefits) by the PW (costs), one obtains the relationship for B/C ratio analysis:

$$PW \text{ (benefits)}/PW \text{ (costs)} \geq 1. \tag{13.3}$$

Equation (13.3) illustrates the standard version of the B/C ratio. An example of the B/C ratio from the data in Table 13.1 using the CFAT values with no rate of return, what is the B/C ratio for the cash flows after taxes for Project A? (The values are listed in Table 13.3.)

The B/C ratio using a zero discount rate starts with the present worth of the benefits (positive cash flows) in the cash flow after taxes which is:

$$PW \text{ (Benefits)} = 11,424 + 13,406 + 7,212 + 13,482 + 12,000 = 57,524.$$

Present worth of the costs (negative cash flows) in the cash flow after tax costs is:

$$PW \text{ (Costs)} = 30,000.$$

Benefit-Cost Ratio $=$ PW (Benefits)$/$PW (Costs) $= 57{,}524/30{,}000 = 1.92 > 1.0$.

Therefore, at a zero rate of return, the benefit/cost ratio exceeds one and the project is acceptable.

If the MARR is 15%, the B/C ratio can be determined using the data of Table 13.1.

The present worth of the discounted benefits (positive cash flows) in the cash flow after taxes is:

$$PW \text{ (Benefits)} = 9{,}934 + 10{,}137 + 4{,}742 + 7{,}708 + 5{,}966 = 38{,}487.$$

The present worth of the costs (negative cash flows) in the cash flow after tax costs is:

$$PW \text{ (Costs)} = 30{,}000.$$

Benefit-Cost Ratio $=$ PW (Benefits)$/$PW (Costs) $= 38{,}487/30{,}000 = 1.28 > 1.0$.

Thus, the effect of a high MARR will generally greatly reduce the B/C ratio and could make it negative and the project unacceptable.

When comparing alternatives, the study period must be the same when using present worth values. The alternative with the highest benefit/cost ratio that is greater than 1.0 is the best acceptable alternative project. If the project all have negative B/C ratios the one closest to 1.0 is the best if a project must be selected; but it does have a negative present worth. If discounted values are used and the ratio is greater than 1.0, the project will also meet the MARR requirements. If the projects have unequal lives, then the B/C should be calculated on an annual worth basis.

In comparing alternatives, projects with large investments may have a higher present worth value than a lower investment project, but the B/C ratio may favor the lower present worth project.

13.4.2 TRAFFIC INTERSECTION EVALUATION

This is a different version of the problem presented in Strategic Cost Analysis [4]. This problem considers a traffic light vs. a traffic circle proposed for a dangerous intersection. In the past 3 years an average of 2 fatalities and 6 major accidents have occurred. The installation of the traffic light signal will cost $500,000, the traffic circle will cost $4,000,000. The annual maintenance for the traffic circle will be $10,000 and the annual maintenance of the signal will be $20,000. The safety engineers expect the fatalities using the traffic light to be reduced by 20% and the serious injuries to be reduced by 40%. They also predict that the traffic circle will reduce fatalities will be reduced by 80% and serious injuries reduced by 50%. The cost of a fatality is estimated at $5,000,000 and a serious injury at $200,000. The discount rate for such projects is 5% (funded by bonds) and the project life is estimated to be 25 years for both alternatives. What is the B/C ratio for each of the alternatives.

Present Worth Values are:

Traffic Light Study

Costs

Investment $= $ $500,000

PW (Annual Maintenance) $= $20,000 (P/A, i = 5\%, n = 25)$

$= $20,000 [(1.05)^{25} - 1]/[0.05(1.05)^{25}] = $20,000 (14.094)$ $= $ $281,880

Total Costs $= $ $781,890

Benefits

PW (Fatality Savings) $= (0.20) \times 5,000,000$

$(P/A, i = 5\%, n = 25) = $1,000,000 \times 14.094$ $= $ $14,094,000

PW (Injury Saving) $= (0.40) \times 200,000$

$(P/A, i = 5\%, n = 25) = 80,000 \times 14.094$ $= $ $1,127,520

Total Benefits $= $ $15,221,520

Benefit/Cost Ratio $= 15,221,520/781,890 = 19.47.$

Traffic Circle Study

Costs

Investment $= $ $4,000,000

PW (Annual Maintenance) $= $10,000 (P/A, i = 5\%, n = 25)$

$= $20,000 [(1.05)^{25} - 1]/[0.05(1.05)^{25}] = $10,000 (14.094)$ $= $ $140,940

Total Costs $= $ $4,140,940

Benefits

PW (Fatality Savings) $= (0.80) \times 5,000,000)$

$(P/A, i = 5\%, n = 25) = $4,000,000 \times 14.094$ $= $ $56,376,000

PW (Injury Saving) $= (0.50) \times 200,000$

$(P/A, i = 5\%, n = 25) = 100,000 \times 14.094$ $= $ $1,409,400

Total Benefits $= $ $57,785,400

Benefit/Cost Ratio $= 57,785,400/4,140,940 = 13.95.$

The present worth of both proposals are much greater than 1.0, but the traffic light is preferred as its B/C ratio is the largest of the two. However, if one compares the projects on a present worth basis, the traffic circle would be preferred.

13.5 MODIFIED BENEFIT/COST RATIO

The M B/C considers the net benefits divided by the total investment costs. The net benefits are the present worth of all the annual benefits minus the present worth of all the annual operating costs and cost term is only the present worth of all the investment costs, which usually occur at time zero but may also be in other periods for large construction projects. This method results in a higher ratio value, but generally does not alter the preference between the alternatives. If the projects have identical project lives, the present worth approach values can be used, but if the project lives are different the average annual values should be use. The values will be higher than the conventional B/C ratio, but the relationship with respect to 1 as to being greater or less than one will not change. It results in a higher B/C number, which makes the project appear better. The investment costs may occur over more than time zero and it would include the present worth of all investment costs:

Modified Benefit/Cost $=$ (Benefits $-$ Operating Costs)/Total Investment Costs.   (13.4)

If one takes the data for the Traffic Light vs. Traffic Circle problem, one has the following.

Traffic Light Study
Costs

Investment	$=$	$500,000
PW (Annual Maintenance) $= \$20,000 \, (P/A, i = 5\%, n = 25)$		
$= \$20,000 \, [(1.05)^{25} - 1]/[0.05(1.05)^{25}] = \$20,000(14.094)$	$=$	$281,880
Total Costs	$=$	$781,880

Benefits

PW (Fatality Savings) $= (0.20) \times 5,000,000$		
$(P/A, i = 5\%, n = 25) = \$1,000,000 \times 14.094$	$=$	$14,094,000
PW (Injury Saving) $= (0.40) \times 200,000$		
$(P/A, i = 5\%, n = 25) = 80,000 \times 14.094$	$=$	$1,127,520
Total Benefits	$=$	$15,221,520

Modified Benefit/Cost Ratio $= (15,221,520 - 281,890)/500,000 = 14,939,641/500,000 = 29.88$.

Traffic Circle Study
Costs

Investment	$=$	$4,000,000
PW (Annual Maintenance) $= \$10,000 \, (P/A, i = 5\%, n = 25)$		
$= \$20,000 \, [(1.05)^{25} - 1]/[0.05(1.05)^{25}] = \$10,000(14.094)$	$=$	$140,940
Total Costs	$=$	$4,140,940

Benefits

$$PW \text{ (Fatality Savings)} = (0.80) \times 5,000,000$$
$$(P/A, i = 5\%, n = 25) = \$4,000,000 \times 14.09 \qquad\qquad = \$56,376,000$$
$$PW \text{ (Injury Saving)} = (0.50) \times 200,000$$
$$(P/A, i = 5\%, n = 25) \times 100,000 \times 14.094 \qquad\qquad = \underline{\$1,409,400}$$

$$\text{Total Benefits} = \$57,785,400$$

Modified Benefit/Cost Ratio $= (57,785,400 - 140,940)/4,000,000 = (57,664,460/4,000,000)$
$= 14.41$.

Both ratios increased, but the traffic light ratio increased more and has the highest ratio. The modified B/C ratio will always give a higher value, but it will not change the alternative selection. The higher value may cause people think it is better, but the changes in the numerator and denominator are the same and since the numerator is always larger for the ratio to be greater than one, the ratio will increase under the M B/C ratio.

The calculations could also be performed on an annual worth cost basis by converting the investment costs to an annual worth (AW) by using $(A/P, i, n)$ and using the original costs and will be illustrated for the traffic light.

Traffic Light Study
Costs

AW (Annual Investment)
$$= \$500,000 \ (A/P, i = 5\%, n = 25)$$
$$= 500,000 \ ((0.05(1 + 0.05)^{25})/((1 + 0.05)^{25} - 1))$$
$$= \$500,000 \ (0.070952) \qquad\qquad = \$35,476$$
$$AW \text{ (Annual Maintenance)} = \$20,000 \qquad\qquad = \underline{\$20,000}$$

$$\text{Total Annual Costs} = \$55,476$$

Benefits

$$AW \text{ (Fatality Savings)} = (0.20) \times 5,000,000 \qquad\qquad = \$1,000,000$$
$$AW \text{ (Injury Saving)} = (0.40) \times 200,000 \qquad\qquad = \underline{\$80,000}$$

$$\text{Total Annual Benefits} = \$1,080,000$$

Modified Benefit/Cost $= (\$1,080.000 - 20,000)/(\$35,476) = 29.88$ (Annual Costs).

The M B/C is the same whether on a present worth basis or on an annual cost basis. In many instances there are several annual costs and it is easier to use the annual cost basis. If one uses annual costs, one must convert the initial investment cost to an annual investment cost.

13.6 POSITIVE AND NEGATIVE PROJECT BALANCES

13.6.1 INTRODUCTION

A Project Balance (PB) determines the project balance at the end of each period. It usually starts with a negative balance, which is a result of the initial investment. An excellent reference for project balances is that by Chan and Sharp-Bette [2]. The balances are summed and the positive future balances reduce the initial negative investment. It is a future worth approach being accumulated from the starting period to the end of the project. The balance lasts through the period and then the next value is added at the end of the period. Each period, usually a year, changes at the end of the period and the last period has no area. This procedure permits one to select between projects which may have similar present worth evaluations. Those projects which have large positive cash flows in the initial periods will have a higher project balance. This method is to be used for comparing projects which have the same study period. The ratio of the negative area balance to the positive area balance is an indication of the risk of the project as the higher the ratio, the higher the risk.

13.6.2 PROJECT A EXAMPLE PROBLEM

If one takes Project A, from Table 13.1 the future worth of the project would be:

$$FW(CFAT) - 15\%)$$
$$= 30,000(1.15)^5 + 11,424(1 + 0.15)^4 + 13,406(1 + 0.15)^3$$
$$\quad + 7,212(1 + 0.15)^2 + 13,482(1 + 0.15) + 12,000$$
$$= -60,341 + 19,980 + 20,389 + 9.538 + 15,504 + 12,000$$
$$= \$17,070$$

(if > 0.0 accept if single alternative or if greatest value among alternatives)

and

$$PW(CFAT - 15\%) = 17,170/(1.15)^5 = 8,487.$$

The project balance approach calculates the project balance at the end of each year and time zero is also included. The future worth of the initial period balance is added to end period balance to create the initial period balance for the next period. The balance is added to the previous balance to determine the end of the current years cumulative cash balance. The last value of the cumulative balance is identical to the future worth of the cash flows for the project. The area represented for the current project balance times the period length, which is one for all periods. The last value which has a zero length as it is the end of the project and does not contribute to the area balance, but is part of the cumulative balance. The results for Project A are shown in Table 13.4 and Figure 13.2 and the large negative areas represent an indication of the risk of the project.

Table 13.4: Project balance Project A calculations for cumulative balances and area balances

Project Balance (PB) for Project A	Cumulative Balance		Area Calculation	Area Balance	
	(-) Balance	(+) Balance		Negative Area	Positive Area
PB0 = -30,000	-30,000		-30,000 *1	-30,000	
PB1 = -30,000*(1 + 0.15) + 11,424	-23,076		-23,076 *1	-23,076	
PB2 = -23,076*(1 + 0.15) + 13,406	-13,131		- 13,131 *1	-13,131	
PB3 = -13,131*(1 + 0.15) + 7,212	- 7,889		- 7,889 *1	- 7,889	
PB4 = - 7,889*(1 + 0.15) + 13,482		+ 4,409	+ 4,409 *1		+ 4,409
PB5 = 4,409*(1+ 0.15) +12,000		+ 17,070	+17,070 *0		0
Total				-74,096	+ 4,409

*Note that the final positive project balance is the same as the future worth of the project.

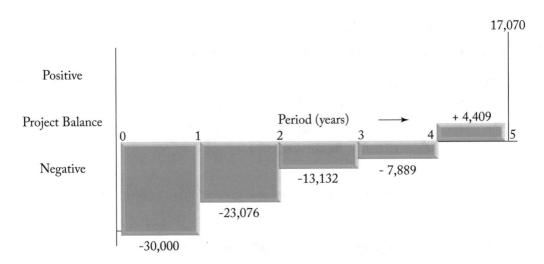

Figure 13.2: Project balance diagram for Project A example problem.

The total negative area of the project is −$74,097 and the positive area is +$4,409 so the project is a somewhat "risky" project. The payback year using the Project Balance method would not occur until the end-of-period 4 and the discounted cash flows also would be in the 4th year, but the undiscounted total cash flow analysis would be the end-of-period 3. In general, the project balance method is more negative as the initial cash flow is throughout the study period and then increased by the MARR. Projects with longer study periods should have lower ratios of negative area/positive area than projects of shorter study periods.

The ratio of Negative Area to the Positive Area is:

$$Negative\ Area/Positive\ Area = 74{,}097/4{,}409 = 16.81.$$

13.6.3 PROJECT Z EXAMPLE PROBLEM

Project (Z) with an initial cost of $30,000 and a set of annual net cash flow after taxes for 5 years to be − $10,000, + $8,000, + $15,000, + $25,000, and + $ 34,147.

The return rate is 15% and the present worth would be:

$$PW(CFAT - 15\%)$$
$$= -30{,}000 - 10{,}000/(1 + 0.15) + 8{,}000/(1 + 0.15)^2$$
$$+ 15{,}000/(1 + 0.15)^3 + 25{,}000/(1 + 0.15)^4 + 34{,}147/(1 + 0.15)^5$$
$$= -30{,}000 - 8.696 + 6{,}049 + 9{,}863 + 14{,}294 + 16{,}077$$
$$= \$8{,}487.$$

This present worth is identical to Project A as calculated in Table 13.1. The project balance, however, is different as shown by Table 13.5 and Figure 13.3. The future worth would also be the same at $17,070.

Table 13.5: Project balance Project Z calculations for cumulative balances and area balances

Project Balance (PB) for Project A	Cumulative Balance		Area Calculation	Area Balance	
	(-) Balance	(+) Balance		Negative Area	Positive Area
PB0 = -30,000	-30,000		-30,000 *1		
				-30,000	

The negative project balance for Project Z, at − $167,175, is much more negative than the project balance for Project A, at − $73,766. and indicates a higher risk (more than double) involved with Project Z. Although the present worth values are the same over the same project lives, the risk for loss is much greater for Project Z than for Project A. There is a positive cumulative balance for Project A at the end-of-period 4, but the Project Z becomes positive only at the end of the project in period 5. Since there is no positive are for Project Z, the ratio of the negative area to positive area would be infinity and this indicates a highly risky project as a positive cumulative balance first occurs at the end of the last period.

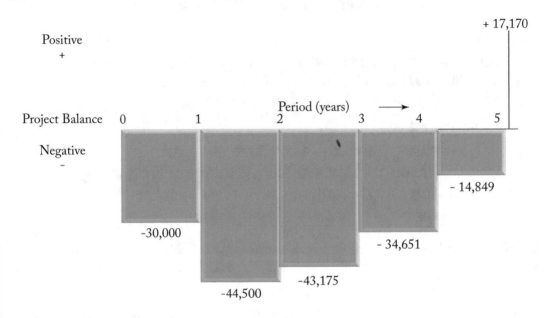

Figure 13.3: Project balance diagram for Project Z example problem.

13.7 SUMMARY

Five additional methods for the evaluation of projects have been presented—the internal rate of return (IRR), the modified internal rate of return (MIRR), the benefit/cost ratio (B/C), the modified benefit cost ratio (M B/C), and the project balance approach (PB). The IRR determines the rate of return at which the present worth becomes zero using a trial and error approach. The MIRR was an estimate of the IRR and will be greater than the MARR, but less than the IRR and is being replaced by directly calculating the IRR. The B/C ratio is the ratio of the positive cash flows to the total of the negative cash flows and the investment. The M B/C ratio considers the ratio of the net benefits (positive cash flows minus the negative cash flows) to the investment and will result in a higher ratio value, but generally will not change the selection between projects with the same study period. The project balance approach is used to compare acceptable projects with present worth that are similar to look at the risk with respect when the positive cash flows occur. It is based upon future worth calculations of projects with equivalent future worth values and equal project lives, and projects with the greater net negative cash flow areas are the more risky projects. Projects with short study periods, large investments, and high MARR values will have large negative areas even though large revenues may occur in the latter project periods.

13.8   REFERENCES

[1] Creese, Robert C. and Adithan, M., *Strategic Cost Analysis for Project Managers and Engineers*, New Academic Science Limited, Tunbridge Wells, UK, pp. 83–108, 2012. 197

[2] Park, Chan S. and Sharp-Bette, Gunter P., *Advanced Engineering Economics*, John Wiley & Sons, Inc., New York, pp. 207–209, 231–236, and 246–253, 1990. 207

[3] Newnan, Donald G., Eschenbach, Ted G., and Lavelle, Jerome P., *Engineering Economic Analysis*, 11th ed., Oxford University Press, New York, p. 655, 2012. 197

[4] Creese, Robert C. and Adithan, M., *Strategic Cost Analysis for Project Managers and Engineers*, New Academic Science Limited, Tunbridge Wells, UK, pp. 95–105, 2012. 203

13.9   EVALUATIVE QUESTIONS

1. Two alternative public works projects to prevent flooding by hurricanes are under consideration. If the MARR is 4%, which of the projects should be selected? Use both the conventional and modified B/C ratios.

	Project HC	Project LC
Capital Investment for Dams and Pumps	$9,000,000	$7,500,000
Annual Operations and Maintenance	$250,000	$200,000
Annual Benefit	$400,000	$300,000
Useful Project Life (years)	60	45

2. Calculate the B/C ratio and the M B/C ratio for the traffic circle and traffic light using annual worth values. Also determine the *PW* values for both the traffic light and traffic circle.

3. Use a value of $7,000,000 for a saved life and calculate the new B/C ratio for the traffic circle and traffic light.

4. An investment of $800,000 was made for a new process which is expected to generate an annual revenue of $350,000 with annual expenses of $100,000 for 7 years before being replaced. The equipment has a MACRS-GDS of 5 years and the MARR is 15% and the data is in Table 13.6. You may want to add additional columns to determine the answers.

   (a) Determine the Internal Rate of Return (IRR).

   (b) Determine the MIRR.

   (c) Determine the B/C ratio on a before tax and depreciation basis (only revenues, expenses, and investment).

Table 13.6: Data for Problem 4

Year	Revenues	Expenses	Net CFBT	Depreciation Percent	Depreciation Amount	Taxable Income	Taxes Paid	Net Profits	CFAT	Cumulative CFAT	Discounted CFAT	Cumulative Discounted CFAT
0	0	800,000	-800,000	0	0	0	0	0	-800,000	-800,000	-800,000	-800,000
1	350,000	100,000	250,000	20.00	160,000	90,000	36,000	54,000	214,000	-586,000	186,087	-613,913
2	350,000	100,000	250,000	32,00	256,000	-6,000	-2,400	-3,600	252,400	-333,600	190,851	-423,062
3	350,000	100,000	250,000	19.20	153,600	96,400	38,560	57,840	211,440	-122,160	139,025	-284,037
4	350,000	100,000	250,000	11.52	92,160	157,840	63,136	94,704	186,864	64,704	106,840	-177,197
5	350,000	100,000	250,000	11.52	92,160	157,480	63,136	94,704	186,864	251,568	92,904	-84,293
6	350,000	100,000	250,000	5.76	46,080	203,920	82,568	122,352	168,432	420,000	72,818	-11,475
7	350,000	100,000	250,000	0	0	250,000	100,000	150,000	150,000	570,000	56,391	44,916
Totals	2,450,000	1,500,000	950,000	100	800,000	950,000	380,000	570,000	570,000		44,916	

(d) Determine the M B/C ratio on a before tax basis and depreciation basis.

(e) Calculate the FW of the CFAT.

(f) Prepare a Project Balance Diagram on the CFAT and determine the total negative area and the total positive areas.

(g) Determine the book value of the equipment over the project life, starting at year zero.

(h) What is the payback period using CFAT and with using discounted CFAT?

(i) Calculate the ROI and ROI-D, AW (ROI), and AW-b (ROI).

5. Rework Problem 4 using straight line depreciation (MACRS-ADS) for 5 years with mid-year depreciation.

6. Rework Problem 4 using a MARR of 5% instead of 15%.

7. Make a project balance diagram for Problem 4. Calculate the future worth of the project and use that as a check for your calculations.

CHAPTER 14

Introduction to Risk Analysis

14.1 INTRODUCTION

The basic methods of risk analysis involve varying selected input parameters of the model to determine their effect upon the cash flows. The probabilistic methods for estimating risk analysis are more advanced and presented in the following chapter. The positive and negative project balances of the previous chapter can also be used for an approximate conservative estimate of project risk and more realistic methods will be presented.

The projects considered involve data for future events such as revenues, expenses, depreciation rates and methods, taxes, and a desired rate of return. The values used are the best estimates that are available when the project study is made and the future is full of uncertainty. The longer the project, the higher degree of uncertainty in the future data. In the previous chapters the data were point estimates of the variables that were assumed known with certainty, but one should consider variation to determine which of the variables are most critical and the sensitivity of the parameter to change. In this chapter, discrete changes will be considered to estimate the variation to estimate project risk.

14.1.1 RISK VS. UNCERTAINTY

Risk refers to situations which can be described by some outcomes whose probabilities can be estimated. These probabilities can be discrete or continuous and the distributions, and parameters are assumed to be known.

Uncertainty implies the probabilities, distributions, and/or parameters are not known. The techniques for considering decisions under conditions of uncertainty are more advanced than the scope of this book. More detailed discussions on the differences between risk and uncertainty are presented in the reference [1]. Thus, the focus in this and the following chapter will be risk analysis [2] but the variability in the risk analysis is often called the uncertainty of the project. Two basic approaches for considering risk presented in this chapter are:

1. Sensitivity Analysis and

2. Optimistic-Pessimistic Analysis (Scenario Analysis)

These two techniques will be illustrated by several examples to emphasize the methodology for obtaining results. The results are used to illustrate the effect of variation of the input variables upon the output.

14.2   SENSITIVITY ANALYSIS

Sensitivity analysis is an approach for examining the impact of change of selected critical parameters in the estimate. The present worth method is frequently used to evaluate the percentage change in one variable while the other variables remain fixed. The variables which will be examined for change are selling price, capacity utilization, investment life, return rate, and total cost changes for an example problem concerning a 3D rapid prototyping project for tooling production.

14.2.1   INNOVATIVE 3D RAPID PROTOTYPING AND TOOLING CENTER EXAMPLE PROBLEM

A group of investors are planning to start a 3D rapid prototyping and tooling center to provide tooling and prototypes for the various manufacturing companies in Manufacturing Valley. The plant initial investment will be $30 million with $10 million for the physical plant construction and $20 million for installation and equipment. The plant will have 50 engineers and technicians with an average salary of $60,000/year and a management and sales force of 6 employees with an average salary of $80,000/year. The process uses a wire feed which is melted by a laser source.

The planned processing capacity is 100 kg of wire per hour with an effective product yield of 80% and the other 20% represents scrap and test specimens. The maximum capacity would be 125 kg per hour. The life of the facility is estimated to be 10 years for investment recovery. The operating costs are estimated at $20/hour, the annual equipment investment is $1 million, the annual depreciation at $2 million, and the total taxes are estimated at 25%. The plant operation is 24 hours/day and for 340 days per year and 25 days for shutdowns and holidays and the annual utility costs are $200,000. The sales revenue is expected to be $40/kg of product sold and the wire cost is expected to be $10/kg used. The data for analysis is in Table 14.1.

The present worth of the investment at the 15% MARR, that is $PWI(15\%)$:

$$PWI(15\%) = \$-30,000,000 + \$8,454,000 \times [P/A, i = 15, n = 10]$$
$$= \$-30,000,000 + \$8,454,000 \times 5.0188$$
$$= \$+12,428,900.$$

The calculated IRR of 25.21% is greater than 15% MARR and the project is approved for further study. The next steps are to investigate the sensitivity of the $PWI(15\%)$ to changes in selling price, capacity utilization, tax rate, investment life, return rate, and total cost changes.

14.2.2   SELLING PRICE SENSITIVITY

The selling price (SP) will be evaluated at a 20% decrease, 10% decrease, zero change, 10% increase, and 20% increase which will result in the selling price levels of 32, 36, 40, 44, and

Table 14.1: Innovative 3D rapid prototyping and tooling center

Land and Building Construction	= $10,000,000
Equipment and Controls	= $20,000,000
Total Investment	= $30,000,000
Annual Revenues	
Product Sales	
340 days/yr * 24 hr/day * 100 kg/hr * 0.8 kg product/kg wire used*$ 40/kg	= $26,112,000
Annual Expenses	
Labor	
Engineers and Technicians	
50 Employees @ $60,000	= $3,000,000
Management and Sales	
6 Employees @ $80,000	= $480,000
Operating Costs	
Materials 340 days/yr*24 hr/day*100 kg/hr*1*10 $/kg	= $8,160,000
Annual Utility Costs	= $200,000
Yearly New Equipment Investment	= $1,000,000
Annual Depreciation Expenses	= $2,000,000
Total Annual Expense	$14,840,000
Gross Profit	11,272,000
Taxes @25%	2,818,000
Net Profits	8,454,000

48 $/kg. The expression for revenue becomes:

$$\text{Revenues} = 652{,}800 \times \text{ selling price (SP)}$$
$$= 652{,}800 \times SP. \tag{14.1}$$

Then the present worth of the investment at a 15% MARR, that is $PWI(15\%)$:

$$
\begin{aligned}
PWI(15\%) = {} & \text{Investment} + (\text{Revenues} - \text{Expenses}) \times (1 - \text{Taxrate (decimal)}) \\
& \times (P/A, i = 15, n = 10) \\
= {} & -30{,}000{,}000 + (652{,}800 \times SP - 14{,}840{,}000) \times (1 - 0.25) \times 5.0188 \\
= {} & -30{,}000{,}000 + 2{,}457{,}204 \times SP - 55{,}859{,}244 \\
= {} & 2{,}457{,}204 \times SP - 85{,}859{,}244.
\end{aligned}
\tag{14.2}
$$

Solving for various costs and the results are in Table 14.2.

Table 14.2: Selling price sensitivity

Sales Price ($/kg)	Sales Price Change (%)	PWI (15%) ($)
32	-20	-7,228,700
36	-10	2,600,100
40	0	12,428,900
44	+10	22,257,700
48	+20	32,086,500

Thus, one observes that the operation is very sensitive to selling price, as a 10% change has approximately a $10 million change in the present worth of the investment.

14.2.3 PROCESSING CAPACITY SENSITIVITY

The processing capacity (PC) will be evaluated similarly to that of the selling price, but both the revenues and the expenses have components related to the processing capacity. The relation for the revenues will be:

$$\text{Revenues} = 340 \times 24 \times PC \times 0.8 \times 40 = 261,120 \times PC. \tag{14.3}$$

The relation for the material expenses will be similar and is

$$\text{Material Expenses} = 340 \times 24 \times PC \times 1 \times 10 = 81,600 \times PC.$$

This will change the total expense expression to:

$$\text{Expenses} = 81,600 \times PC + (14,840,000 - 6,680,000).$$
$$= 81,600 \times PC + 6,680,000 \tag{14.4}$$

$$PW(15\%) = \text{Investment} + (\text{Revenues} - \text{Expenses}) \times (1 - \text{Taxrate(decimal)})$$
$$\times (P/A, i = 15, n = 10)$$
$$= -30,000,000 + (261,120 \times PC - (81,600 \times PC + 6,680,000))$$
$$\times (1 - 0.25) \times 5.0188$$
$$= -30,000,000 + (675,731 \times PC - 25,144,188)$$
$$= 675,731 \times PC - 55,144,188. \tag{14.5}$$

Solving for various capacity levels, the results are in Table 14.3.

Table 14.3: Process capacity sensitivity

Processing Capacity (kg/hr)	Processing Capacity Change (%)	PW (15%) ($)
80	-20	-1,005,700
90	-10	5,671,600
100	0	12,428,900
110	+10	19,186,200
120	+20	25,943,500

The present worth is also very sensitive to the processing capacity (PC) as a 10% change results in a nearly $7 million change. It is not as sensitive as the selling price, but it would be a critical parameter to monitor.

14.2.4 TAX RATE SENSITIVITY

The effect of tax rates (TR) receives considerable attention in the political world and the TR being considered range from 20–40%, and that range will be considered. This results to changes of −20%, base case + 20%, + 40%, and + 60%. The values for the revenues and expenses before taxes would be for the base case:

$$\text{Revenues} = \$26,112,000$$
$$\text{Expenses} = \$14,840,000.$$

$$PW(15\%) = -30,000,000 + (26,112,000 - 14,840,000) \times (1 - TR)$$
$$\times (P/A, i = 0.15, n = 10)$$
$$= -30,000,000 + (11,272,000) \times ((1 - TR) \times 5.0188$$
$$= -30,000,000 + 56,571,914 \times (1 - TR). \tag{14.6}$$

Solving for the effects of taxes on the $PW(15\%)$ results are in Table 14.4.

Even though the tax rate has changed considerably both as the amount applied and the percentage increase, the project still has a positive present worth, $PW(15\%)$, which is greater than 10% of the initial investment. Since the percentage changes are larger than the other comparisons, they cannot be compared directly, but the changes are smaller than one normally would expect. This indicates that the effects of sales and operations management performance have a much greater effect upon the project present worth than the tax rate.

Table 14.4: Tax rate sensitivity

Tax Rate Applied (%)	Tax Rate Change (%)	PWI (15%) ($)
20	-20	15,257,500
25	0	12,428,900
30	+20	9,600,300
35	+40	6,771,700
40	+60	3,943,100

14.2.5 INVESTMENT LIFE SENSITIVITY

The investment life (IL) of the facility will affect the $[P/A, i, n]$ term in the $PW(15\%)$ expression. The various investment lives considered will be 8, 9, 10, 11, and 12 years. The expression would be:

$$
\begin{aligned}
PW(15\%) &= -30,000,000 + (26,112,000 - 14,840,000) \times (1 - 0.25) \\
&\quad \times (P/A, i = 0.15, n = 8, 9, 10, 11, 12) \\
&= -30,000,000 + (11,272,000) \times ((0.75) \\
&\quad \times (P/A, i = 0.15, n = 8, 9, 10, 11, 12) \\
&= -30,000,000 + 8,454,000) \times (P/A, i = 0.15, n = 8, 9, 10, 11, 12). \quad (14.7)
\end{aligned}
$$

Using this expression and varying the investment life from 8–12 years with the results in Table 14.5.

Table 14.5: Investment life sensitivity

Investment Life (IL) (years)	Percent Change (%)	P/A, i = 0.15 n = IL	PW (15%) ($)
8	-20	4.4972	8,016,300
9	-10	4.7716	10,339,100
10	0	5.0188	12,428,900
11	10	5.2337	14,245,700
12	20	5.4206	15,825,800

One notices that the operations sensitivity to investment life to a 10% change in invest-ment life changes the present worth, $PW(15\%)$, by roughly 2 million dollars compared to the much greater 7 and 10 million dollar changes by sales price and processing capacity changes.

14.2.6 REQUIRED RATE OF RETURN SENSITIVITY

The MARR of the facility will affect the $[P/A, i, n]$ term in the PW (MARR)) expression. The various required return values considered will be 12, 13.5, 15, 16.5, and 18% which represent changes of -20, -10, 0, $+10$, and $+20\%$ of the initial MARR. The expression would be:

$$PW(15\%) = -30,000,000 + (26,112,000 - 14,840,000) \times (1 - 0.25)$$
$$\times (P/A, i = 0.12, 0.135, 0.15, 0.165, 0.18, n = 10)$$
$$= -30,000,000 + (11,272,000) \times ((0.75)$$
$$\times (P/A, i = 0.12, 0.135, 0.15, 0.165, 0.18, n = 10)$$
$$= -30,000,000 + 8,454,000)$$
$$\times (P/A, i = 0.12, 0.135, 0.15, 0.165, 0.18, n = 10). \tag{14.8}$$

Using this expression and varying the MARR from 12–18% with the results in Table 14.6.

Table 14.6: MARR sensitivity

MARR (i) Percent (%)	Percent Change of MARR	P/A, i = MARR n = 10	PW (MARR(i)) ($)
12	-20	5.6502	16,513,900
13.5	-10	5.3195	14,971,100
15	0	5.0188	12,428,900
16.5	10	4.7446	10,110,800
18	20	4.4941	7,993,100

The magnitude of the PW change in the MARR is similar to that of the investment life, but as the MARR requirement increases the PW decreased. The change per increase was less than 2 million per 10% change which is similar to the changes in investment life, but in the opposite directions.

14.2.7 TOTAL COST SENSITIVITY

The total cost (TC) can often change faster than the revenues, so an examination of similar changes in the total costs as was done for changes in the total revenues will be presented. An adjustment factor (AF) will be used on the expenses to have the same percentage changes as occurred in the revenue increases. The cost adjustment factors (AF) will be: 0.80, 0.90, 1.0, 1.1, and 1.2:

$$PWI(15\%) = \text{Investment} + (\text{Revenues-Expenses} \times AF) \times (1 - \text{Taxrate(decimal)}$$
$$\times (P/A, i = 15, n = 10)$$
$$= -30,000,000 + (26,112,000 - 14,840,000 \times AF) \times (1 - 0.25) \times 5.0188$$
$$= -30,000,000 + (26,112,000 - 14,840,000 \times AF) \times 3.7641. \qquad (14.9)$$

Using this expression and the AF values of 0.80, 0.90, 1.0, 1.1, and 1.2, the present worth results are in Table 14.7.

Table 14.7: Total cost sensitivity

Cost Adjustment Factor (AF)	Percent Change in AF (%)	PW (15%) ($)
0.8	-20	23,600,800
0.9	-10	18,014,600
1.0	0	12,428,900
1.1	10	6,843,000
1.2	20	1,257,100

The lower the cost adjustment factor, the higher the present worth. The changes as a result of the changes in total costs are large, but not as large as the sales revenue or the process capacity changes. But it is an area that management primarily controls at the operations area, rather than at the marketing and sales areas, and is the one that production management should focus on.

14.3 OPTIMISTIC-PESSIMISTIC ANALYSIS

Optimistic-Pessimistic Analysis is used to evaluate variation in one or more variables and three different levels of variation for each of the variable(s). Three cases are usually considered for each variable and they are: a "worst-case" (Pessimistic or P), "most-likely case" (Most Likely or ML), and "best-case" (optimistic or O). Often more than one variable is analyzed in the study. The worst case is one which the results would be lower in less than 5 (or 10)% of the cases and the best case would be exceeded in only 5 (or 10)% of the cases. This is a simple and effective method for analyzing the effect of two variables. More variables can be considered, but the tables would be considerably more complex.

14.3.1 INNOVATIVE 3D RAPID PROTOTYPING AND TOOLING CENTER INVESTOR CONCERNS

The investors for the rapid prototyping and tooling center had two major concerns after making the sensitivity analysis of the variables. The investors were concerned with the revenues and the investment life as computerized equipment can become obsolete rapidly as noted by the rapid changes in smartphone capabilities. The following three scenarios were developed for the desired rate of return of 15% after taxes for the two variables of Net Revenues and Investment Life. The investment life values were changed to 6 years and 14 years as the depreciation issues under accelerated depreciation would be over in 6 years and major replacements would definitely be required after 14 years.

Table 14.8: Innovative rapid prototyping and tooling center analysis

	Optimistic (O)	Scenarios Most-Likely (ML)	Pessimistic (P)
Capital Investment (Million $ Units)	30	30	30
Investment Life (yrs)	14	10	6
Net Annual Revenues (Million $ Units)	12.40	8.45	4.54

The present worth of three scenarios of the nine total scenarios were evaluated as:

$$PW - O(15\%) = -30 + 12.40[P/A, i = 15, n = 14]$$
$$= -30 + 12.40[5.7245] = 41.0$$
$$PW - ML(15\%) = -30 + 8.45[P/A, i = 15, n = 10]$$
$$= -30 + 8.45[5.0188] = 12.4$$
$$PW - P(15\%) = -30 + 4.54[P/A, i = 15, n = 6]$$
$$= -30 + 4.54[3.7845] = -12.8.$$

Table 14.9 contains the results of all nine scenarios and the reader should check the other calculated values in the solution matrix.

These three scenarios give an simple average of $11.0 million and a range of $53.8 million. The question is which of the two variables is the critical variable, revenue, or investment life. This analysis indicates that the low revenues are the primary cause and not the short life. Only when the revenues are low does the project lose money, and it does so at all three lives considered. This indicates to management that revenues are key and perhaps should be re-examined. The overall average is $11.0 million on equal weight basis compared to the most likely value of 12.4.

Table 14.9: Innovative rapid prototyping and tooling center analysis solution matrix

		Investment Life (years)		
		6	10	14
Net	$12.40	16.9	32.2	41.0
Revenues	$ 8.45	2.0	12.4	18.4
(Million $ Units)	$ 4.54	-12.8	-7.2	-4.0

14.4 SUMMARY

Two commonly used methods are applied for estimating variability and risk are sensitivity analysis and optimistic-pessimistic analysis to indicate which are the critical variables and which are the non-critical variables. These methods utilize the present worth approach and probabilities are not considered. Sensitivity analysis takes only one variable for consideration at a time, but it is straight forward process. The optimistic-pessimistic analysis considers two variables at a time and often indicates which of the two variables is more critical easier than the individual variable analysis. The consideration of three variables is much more difficult than two variables in the optimistic-pessimistic analysis.

14.5 REFERENCES

[1] Garvey, Paul R., *Probability Methods for Cost Uncertainty Analysis*, Marcel Dekker, Inc, New York, p. 27 and p. 338, 2000. 215

[2] Creese, Robert C. and Adithan, M., *Strategic Cost Analysis for Project Managers and Engineers*, New Academic Science Limited, Tunbridge Wells, UK, pp. 109–114, 2012. 215

14.6 EVALUATIVE QUESTIONS

1. The salaries were too low in the data of Table 14.1 and the engineers and technicians were increased to $80,000 and the management and sales were increased to $100,000. The tax rate was decreased to 20%. Since cash flows were was very sensitive to selling price, calculate the present worth of the cash flows and determine the sensitivity of it similar to that of Table 14.2.

2. Two projects are being considered, one with high risk (risky) with a higher investment but higher returns and a more conservative project (traditional). Should the risky project be selected?

Project Challenge

	T (Traditional Project)	R (Risky Project)
Net Investment ($)	150,000	225,000
Total Revenues	80,000	110,000
Total Costs	20,000	20,000
Net Revenues ($)	60,000	90,000
MARR (%)	15	15
Project Life	4	4

(a) Using *PW* analysis, which is the better project? Note that the net investment is 50% higher as well as the net revenues. What are the average annual worth of the investments?

(b) Let the investment life of the traditional project be 5 years instead of 4? How does that alter the selection? Use average annual worth techniques for consideration.

(c) Use the investment live of 4 years for both projects, but the risky project must earn 20% MARR and the traditional project remains at 15%.

(d) What is the B/C ratio for the two projects over 4 years? (Evaluate the risky project at both MARR values.)

3. Project ABC has a high risk has a high investment but higher returns. Should it be selected with the new data?

   Project ABC

	R (Risky Project)
Net Investment ($)	225,000
Total Revenues ($)	90,000
Total Costs ($)	25,000
Net Revenues ($)	65,000
MARR (%)	15
Project Life (years)	5

(a) Should it be selected?

(b) Do a sensitivity analysis by varying the project life for 3, 4, 5, 6, and 7 years. What is the trend?

(c) Do a sensitivity analysis by varying the MARR from 5, 10, 15, 20, and 25%. What is the trend?

(d) What is the rate of return at which the project has a zero present worth value? What is this rate of return named?

(e) What is the discounted and non-discounted B/C ratio for the initial project?

(f) What is the ROI and ROI-D for the project. (assume total costs includes all expenses including depreciation and taxes)?

4. Calculate the six scenarios that were that were not done for the results in Table 14.9. Give the equations, the values used, and the results for six scenarios.

5. The investors were concerned about with the net revenues and the investment life in new process. The following three scenarios were developed for the desired rate of return of 15% after taxes for the 2 variables of net revenues and investment life. The investment life values were changed to 3, 5, and 7 years as the project is risky.

Project Challenge Data

	Scenarios		
	Optimistic (O)	Most-Likely (ML)	Pessimistic (P)
Capital Investment (Million $ Units)	40	40	40
Investment Life (years)	7	5	3
Net Annual Revenues (Million $ Units)	21	18	15

(a) Determine the present worth of the nine possibilities and form a solution matrix. Discuss which of the variables you consider to be most important for this problem.

(b) Calculate the ROI for the three scenarios: optimistic, most-likely, and pessimistic.

(c) Calculate the ROI for the 3 scenarios using the investment live of 7 years for all three.

C H A P T E R   15

Risk Analysis with Probability Considerations

15.1 PROBABILITY METHODS AND TERMINOLOGY

The traditional risk methods of the previous chapter gave point estimates values for the project of interest, but little indication of the potential range of the results or the probability of a loss on the project. This added information is helpful in making decisions about the selection of a particular project. This chapter will present an introduction to probability considerations in the evaluation of projects, but other references are listed for a more detailed coverage of the topic [1, 2].

The key terms are random variables and probability distributions. A random variable can take on several values which can have a probability that can be determined by the probability distribution for that variable. The random variable can have either discrete values predicted by discrete probability distribution (also called discrete probability density function) or continuous values predicted by a continuous probability distribution (or continuous probability density function). In discrete distributions there are a finite number of values and each value as a distinct probability associated with it. With continuous distributions, the set of variables is not countable and the probability density function does not produce a probability value at a point as in the discrete distribution, but a probability range for between two points which are designated as the lower point, L, and the higher or upper point, U.

The key measures of probability distributions are the mean or "expected value" and the "variance" and "standard deviation" which are used to indicate mean and the possible variation of the mean and the data set. The standard deviation is the square root of the variance and is more commonly used to describe the variation about the mean. The formula for the expected value for the discrete probability distribution, which involves a summation is:

$$E(x) = \mu = \sum_{i=1}^{N} p_i x_i \text{ (discrete case)}, \tag{15.1}$$

where

$E(x)$ = expected value of the variable or mean x
μ = symbol for mean
i = number of outcomes of variable x in discrete case
N = total number of outcomes and N is the last outcome

p_i = probability of specific ith outcome occurring.

The formula for the variance of the discrete probability distribution is:

$$Var(X) = \sigma^2 = \sum_{i=1}^{i=N} [x_i - \mu]^2 p_i$$

$$\sigma^2 = \sum_{i=1}^{i=N} [x_i^2 - 2\mu x_i + \mu^2] p_i$$

$$\sigma^2 = \sum_{i=1}^{i=N} [x_i^2] p_i - 2\mu \sum_{i=1}^{i=N} x_i p_i + \mu^2 \sum_{i=1}^{i=N} p_i$$

$$\sigma^2 = \sum_{i=1}^{i=N} x_i^2 - 2\mu \times \mu_i + \mu^2 \times 1$$

$$\sigma^2 = E(x^2) - \mu^2$$

$$\sigma^2 = E(x^2) - [E(x)]^2, \tag{15.2}$$

where

σ^2 = symbol for the variance, which is the square of the standard deviation
σ = symbol for the standard deviation
$E(x)$ = expected value of the variable or mean x which is μ
μ = symbol for the mean
$E(x^2)$ = expected value of the square of the variable x
i = an outcome of variable x in the discrete case
N = total number of outcomes and N is the last outcome
p_i = probability of specific ith outcome occurring.

For continuous probability density functions (PDFs), the probability of an event x between the lower limit of L and the upper limit U is given by:

$$P(L < x < U) = \int_{L}^{U} f(x) dx. \tag{15.3}$$

When the lower and upper limits are the total range of the distribution of the distribution, the value of P (Lowest Value $\leq x \leq$ Highest Value) will be 1.0. When the values are less than the total range, the probability will be less than 1.0.

The formula for the expected value for the distribution between the, that is the mean μ, for the continuous PDF, which involves an integral, is:

$$E(x) = \mu = \int_{L}^{U} x \times f(x) dx, \tag{15.4}$$

where

$E(x)$ = expected value of the variable or mean x

μ = symbol for mean

$f(x)$ = continuous probability distribution function of variable x

U = upper limit of continuous probability distribution of variable x

L = lower limit of continuous probability distribution of variable x.

The formula for the variance of the continuous probability distribution, which also involves an integral, is:

$$Var(X) = \sigma^2 = \int_L^U [x - E(x)]^2 f(x)dx \text{ (continuous case)}$$

$$\sigma^2 = \int_L^U x^2 f(x)dx - [E(X)]^2.$$

This follows the same procedure as in the discrete case except integrals are used instead of summations and results in:

$$\sigma^2 = E(X^2) - [E(X)]^2. \tag{15.5}$$

Examples will now be presented illustrating the use of discrete probability analysis and then will follow with some continuous probability examples.

15.2 DISCRETE PROBABILITY EXAMPLES

15.2.1 DONNIE THE DEALMAKER

Donnie the Dealmaker has invested in a product where he has arranged for three suppliers and has three major customers. He needs all the suppliers and the customer demand is high, and they are completely independent of each other. The information on the costs, revenues proportion of the total supplied by each supplier, and their cost as well as the proportion of the total sold to each customer is in Table 15.1. This problem follows a procedure illustrated by Garvey [3]. All of the product supplied by the suppliers will be sold to the customers.

Table 15.2 shows the revenues, revenue probabilities, costs, cost probabilities, profit amounts profit probabilities, expected profits, and $E(x)^2$. Donnie the Dealmaker wants to know what he will make as a profit and the standard deviation.

The total of the expected profit is 11.0 and that is the expected profit with the suppliers and customers and the probabilities with no inventory or shortage problems. The variance and standard deviation can be found using the last two columns of Table 15.2 and Equation (15.2):

$$\sigma^2 = E(X^2) - [E(X)]^2$$
$$= 206 - [11]^2$$
$$= 85.$$

Table 15.1: Supplier and customer share of product and prices

Supplier	A	B	C
Supplier Share of Product	0.2	0.5	0.3
Supplier Price to Dealmaker	30	40	50
Customer	X	Y	Z
Customer Share of Product	0.1	0.6	0.3
Customer Price Paid to Dealmaker	40	50	60

Table 15.2: Revenues, costs, and profits for Donnie the Dealmaker's product investment

Revenue Amount	Revenue Probability	Cost Amount	Cost Probability	Profit Amount	Profit Probability	Expected Profit	$E(X^2)$
40	0.1	30	0.2	10	0.02	0.2	2
40	0.1	40	0.5	0	0.05	0.0	0
40	0.1	50	0.3	-10	0.03	-0.3	3
50	0.6	30	0.2	20	0.12	2.4	48
50	0.6	40	0.3	10	0.30	3.0	30
50	0.6	50	0.5	0	0.18	0	0
60	0.3	30	0.2	30	0.06	1.8	54
60	0.3	40	0.3	20	0.15	3.0	60
60	0.3	50	0.5	10	0.09	0.9	9
Totals					1.00	11.0	206

The standard deviation is the square root of the variance and is:

$$\sigma = 85^{1/2}$$
$$\sigma = 9.22.$$

Dealmaker Donnie will make an average of $11 in sales, but the profit on any individual sale can range from −$10 to +$30 in $10 increments. The probability of a loss occurs on only one event which has a 3% probability (0.03 in Table 15.2) of occurring.

15.2.2 THE INNOVATIVE 3D RAPID PROTOTYPING AND TOOLING CENTER

The Innovative 3-D Rapid Prototyping and Tooling Center example in Sections 14.2.1 and 14.2.3 had the following capacity values and present worth values as presented in Table 15.3.

The probabilities for each capacity level are as assigned and thus one can determine the expected value (mean), variance, and standard deviation and then determine the mean value of the present worth:

Table 15.3: Innovative 3D rapid prototyping and tooling center data

Processing Capacity Kg/hr	Probability of Event "x"	Present Worth (15%)	Expected Capacity Kg/hr	
x	p(x)	PW/10^3	x*p(x)	x^2 * p(x)
80	0.20	-1,006	16	1,280
90	0.20	5,672	18	1,620
100	0.30	12,429	30	3,000
110	0.20	19,186	22	2,420
120	0.10	25,944	12	1,440
Totals	1.00		98	9,760

$\mu = \text{Mean} = E(x) = \sum x \times p(x) = 98$

$E(x^2) = \sum x^2 \times p(x) = 9{,}760$

$\sigma^2 = \text{Variance} = E(x^2 - (E(x))^2 = 9{,}760 - 98^2 = 9{,}760 - 9{,}604 = 156$

$\sigma = \text{Standard Deviation} = \text{Square root of variance} = 12.49$

The expected value (mean) of the processing capacity is 98 kg/hr, which is less than the designed operating capacity of 100, the variance is 156 (kg/hr)^2, and the standard deviation is 12.49 kg/hr. The present worth as a function of the processing capacity from final version of Equation (14.2) was:

$$PW(15\%) = 675{,}731 \times PC - 55{,}144{,}188. \qquad (14.2)$$

The mean of the present worth would be:

$$\mu(PW(15\%)) = \$675{,}731 \times 98 - 55{,}144{,}188$$
$$= \$11{,}077{,}450.$$

Since the process capacity mean is lower than the base processing capacity value of 100, the present worth value is also lower than its original value of $12,428,900. The standard deviation for the present worth is approximately $8,422,000.

15.3 CONTINUOUS PROBABILITY MODELS

As mentioned previously, the probability at a specific value with a continuous distribution is zero and the probabilities are calculated for a range of values between an lower limit, L and an upper limit, U. The probability is the area between the two limits of the PDF. There are numerous continuous probability density functions and only two will be considered which are the normal distribution and the triangular distribution. The normal distribution is the most commonly used distribution of all the probability density functions, but the triangular distribution is commonly used in estimating and in determining ranges of cost data.

15.3.1 NORMAL DISTRIBUTION PROPERTIES

The normal distribution is frequently assumed for many problems as the central limit theorem indicates that the means from samples are distributed normally. For example, the cost of an item is the total of three items that are independent random variables, then the distribution of the total cost will be normal even if the independent random variables are not normally distributed. Some of the properties about means and variances of distributions are:

$$E\left(\sum_{i=1}^{N} X\right) = E(X1) + E(X2) + \cdots + E(N) \quad \text{for} \quad i = 1, 2, \ldots N \qquad (15.6)$$

$$E(aX1 + bX2) = aE(X1) + bE(X2) \qquad (15.7)$$

$$\text{Variance}\left(\sum_{i=1}^{N} X\right) = \text{Variance } (X1) + \text{Variance } (X2) + \cdots + \text{Variance}(XN) \qquad (15.8)$$

$$\sigma = \text{standard deviation} = \text{Square Root (variance)}$$

$$\text{Standard Deviation } (aX1 + bX1) = [a^2 \text{ variance} - X1 + b^2 \text{ variance} - X2]^{1/2} \qquad (15.9)$$

The cumulative probability for the variable X to a specific value c is given by

$$\text{Probability } (X < c) = \Phi[(c - \mu)/\sigma] = \Phi(Z). \qquad (15.10)$$

Probability values of $\Phi(Z)$ are in Table 15.4.

Basic Normal Probability Examples
Use Equation (15.10) and Table 15.4 to answer the following questions to gain familiarity in obtaining probability values. Table 15.4 is useful, but the NORMSDIST function in Excel® is much better and it, or other equivalent expressions in other software packages, would be easier and faster than look-up tables. If using electronic spreadsheets it would be useful, easier, and more accurate to use the computer function rather than Table 15.4.

   A. If the mean of a normal distribution is 20 and the standard deviation is 10, what is the probability that a random variable selected from that distribution is less than zero?

$$Prob(X < 0) = \Phi[(0 - 20)/10] = \Phi[-2.0] = 0.023 = 2.3\%$$

Table 15.4: Probability values $\Phi(Z)$ for the standard normal distribution Z-Values in 0.05 increments

Z -Value	$\Phi(Z)$	Z -Value	$\Phi(Z)$	Z -Value	$\Phi(Z)$	Z -Value	$\Phi(Z)$	Z -Value	$\Phi(Z)$	Z -Value	$\Phi(Z)$
−3.00	0.001	−2.00	0.023	−1.00	0.159	0.00	0.500	1	0.841	2.00	0.978
−2.95	0.002	−1.95	0.026	−0.95	0.171	0.05	0.520	1.05	0.853	2.05	0.980
−2.90	0.002	−1.90	0.029	−0.90	0.184	0.10	0.540	1.10	0.864	2.10	0.982
−2.85	0.002	−1.85	0.032	−0.85	0.198	0.15	0.560	1.15	0.875	2.15	0.984
−2.80	0.003	−1.80	0.036	−0.80	0.212	0.20	0.579	1.20	0.885	2.20	0.986
−2.75	0.003	−1.75	0.040	−0.75	0.227	0.25	0.599	1.25	0.894	2.25	0.988
−2.70	0.004	−1.70	0.045	−0.70	0.242	0.30	0.618	1.30	0.903	2.30	0.990
−2.65	0.004	−1.65	0.049	−0.65	0.258	0.35	0.637	1.35	0.911	2.35	0.991
−2.60	0.005	−1.60	0.055	−0.60	0.284	0.40	0.655	1.40	0.919	2.40	0.992
−2.55	0.005	−1.55	0.061	−0.55	0.291	0.45	0.674	1.45	0.926	2.45	0.993
−2.50	0.006	−1.50	0.067	−0.50	0.308	0.50	0.691	1.50	0.933	2.50	0.994
−2.45	0.007	−1.45	0.073	−0.45	0.326	0.55	0.701	1.55	0.939	2.55	0.995
−2.40	0.008	−1.40	0.081	−0.40	0.345	0.60	0.726	1.60	0.945	2.60	0.995
−2.35	0.009	−1.35	0.088	−0.35	0.363	0.65	0.742	1.65	0.951	2.65	0.996
−2.30	0.011	−1.30	0.097	−0.30	0.382	0.70	0.758	1.70	0.955	2.70	0.996
−2.25	0.012	−1.25	0.106	−0.25	0.401	0.75	0.773	1.75	0.960	2.75	0.997
−2.20	0.014	−1.20	0.115	−0.20	0.421	0.80	0.788	1.80	0.964	2.80	0.997
−2.15	0.016	−1.15	0.125	−0.15	0.440	0.85	0.802	1.85	0.968	2.85	0.998
−2.10	0.018	−1.10	0.136	−0.10	0.460	0.90	0.816	1.90	0.971	2.90	0.998
−2.05	0.020	−1.05	0.147	−0.05	0.480	0.95	0.829	1.95	0.974	2.95	0.998
−2.00	0.023	−1.00	0.159	0.00	0.500	1.00	0.841	2.00	0.978	3.00	0.999

B. What is the probability that the random variable selected is less than 15?

$$Prob(X < 15) = \Phi[(15 - 20)/10] = \Phi[-0.5] = 0.308 = 30.8\%$$

C. What is the probability that the random variable selected is greater than or equal to 30?

$$Prob(X \geq 30) = 1 - Prob(X < 30) = 1 - \Phi[(30 - 20)/10] = 1 - \Phi[+1.0]$$
$$= 1 - 0.841 = 0.159 \text{ or } 15.9\%$$

D. What is the probability that the random variable selected is between 15 and 30?

$$Prob(15 \leq X \leq 30) = Prob(X \leq 30) - Prob(X < 15)$$
$$= \Phi[(30 - 20)/10] - \Phi[(15 - 20)/10]$$
$$= \Phi(+1.00) - \Phi(-0.50)] = 0.841 - 0.308 = 0.533 = 53.3\%$$

E. If one assumes that Donnie the Dealmaker can assume his data is distributed normally, the probability of a loss would be:

$$Prob(X < 0) = \Phi[(0 - 11)/9.22] = \Phi[-1.193] \approx 0.027 \text{ or } 2.7\%.$$

This is very close to the discrete probability of 3.0%.

This risk of a project is usually considered to be the probability of obtaining a loss, but it can be specified to be a loss at a specific required return.

Cash Flow Normal Distribution Example Problem

The cash flows from a project are presented in Table 15.5. The required rate of return is 15%. What is the present worth of the cash flows on the project with an investment of $40,000, expected mean revenues, and expected standard deviation of the cash flows? What is the probability that the cash flow from the project has a loss?

Table 15.5: Cash flow example data

Year	Expected Cash Flow μ($)	Standard Deviation of Cash Flow σ	Variance of Cash Flow σ^2
0	-40,000	1,000	1,000,000
1	15,000	1,500	2,250.000
2	20,000	2,000	4,000,000
3	20,000	3,000	9,000,000

The first steps are to calculate the mean, variance, and standard deviation of the present worth of the cash flows at the required rate of return. This can be done by:

$$PW(15\%) = -40,000 + 15,000[P/F, 15, 1] + 20,000[P/F, 15, 2] + 20,000[P/F, 15, 3]$$
$$= -40,000 + 15,000 \times 0.8696 + 20,000 \times 0.7561 + 20,000 \times 0.6575$$
$$= \$1,316.$$
$$\mu = PW(15\%) = \$1,316.$$

The variance is calculated by:

$$V(PW) = \sigma(PW)^2 = 1^2 \times 1{,}000^2 + (.8696)^2 \times 1{,}500^2 + (.7561)^2 \times 2{,}000^2 + (.6575)^2 \times 3{,}000^2$$
$$= 1 \times 10^6 + 1.7015 \times 10^6 + 2.2867 \times 10^6 + 3.8908 \times 10^6$$
$$\sigma(PW)^2 = 8.8790 \times 10^6$$
$$\sigma(PW) = 2.980 \times 10^3$$
$$= 2{,}980.$$

$$P(PW < 0) = \Phi[(0 - 1{,}316)/2{,}980] = \Phi[-0.4416] \approx \Phi[-0.45] \approx 0.326 = 0.33 = 33\%$$

This indicates the mean value of the present worth is \$1,316, the standard deviation is \$2,980 and there is a 33% chance that the project will not make the desired return of 15%; that is, it will have a negative present worth value for the desired MARR 33% of the time.

15.3.2 TRIANGULAR DISTRIBUTION PROPERTIES

The triangular distribution is described by a lower limit, the most likely value (which is also called the mode) and an upper limit. Costs of individual items tend to follow the triangular distribution rather than the normal distribution. The equations for the cumulative PDFs, mean, and variance are listed in many sources, such as Wikipedia [4] and Garvey [2].

$$P(x < M) = (x - L)^2/((U - L) \times (M - L)) \quad \text{for} \quad L \le x \le M \qquad (15.11\text{a})$$
$$P(x > M) = 1 - (U - x)^2/((U - L) \times (U - M)) \quad \text{for} \quad M \le x \le L \qquad (15.11\text{b})$$

where
$L = $ Lower limit
$U = $ Upper limit
$M = $ Most Likely Value, also called Mode

$$\mu = E(x) = (L + M + U)/3 \qquad (15.12)$$
$$\sigma^2 = Var(x) = (1/18) \times ((M - L) \times (M - U) + (U - L)^2) \qquad (15.13)$$

Basic Triangular Probability Example
A variable follows the triangular distribution and has a most likely value of 12 and a lower limit of 7 and a upper limit of 20. What is the mean, standard deviation, the probability that the random variable is less than 10, the probability the random variable is between 8 and 10, and

the probability the random variable is greater than 18?

$$\mu = E(x) = (L + M + U)/3 = (7 + 12 + 20)/3 = 13$$
$$\sigma^2 = Var(x) = (1/18) \times ((M - L) \times (M - U) + (U - L)^2)$$
$$= (1/18) \times ((12 - 7) \times (12 - 20) + (20 - 7)^2) = 7.17$$
$$\sigma = (7.17)^{1/2} = 2.68$$
$$P(x < 10) = (10 - 7)^2/((20 - 7) \times (12 - 7) = 0.138 = 13.8\%$$
$$P(8 < x < 10) = P(x < 10) - P(x < 8)$$
$$P(x < 8) = (8 - 7)^2/((20 - 7) \times (12 - 7) = 0.015 = 1.5\%$$

Therefore,

$$P(8 < x < 10) = P(x < 10) - P(x < 8) = 13.8\% - 1.5\% = 12.3\%$$
$$P(X > 18) = 1 - P(x < 18) = 1 - [1 - (20 - 18)^2/((20 - 7) \times (20 - 12))]$$
$$= 1 - [1 - .038] = 0.038 = 3.8\%.$$

PERT and the Cooper and Davidson Approach

The triangular distribution is often used in cost analysis as individual cost components tend to follow the triangular distribution rather than the normal distribution, but the sum of cost components via the central limit theorem will have a normal distribution. The PERT [5] and Cooper and Davidson [6] approaches use modified triangular versions with more weight on the mode than on the outside limits. The Program Evaluation and Review Technique (PERT), utilizes the triangular distribution with a high (optimistic), low (pessimistic), and most likely values and that is similar to the Optimistic-Pessimistic Analysis of Chapter 14. PERT was first used in the late 1950s in the development of the Polaris nuclear submarine program.

As in the use of the Optimistic-Pessimistic Analysis Technique, it is assumed that the high and low values are known (or can be estimated) and are the endpoints of the distribution. The normal distribution with its endpoints of infinity (plus and minus) is not practical in cost analysis situations as negative costs would not be reasonable for a single component. It is very difficult to estimate the end points, and Cooper and Davidson (C&D) have modified the parameters so the end points are 10% values; that is, the low estimate implies there is only a 10% chance of having a value lower than this estimate and the high estimate implies there is only a 10% chance of having a value higher than this estimate. Using this definition, their expressions for the mean and standard deviation are:

$$\mu = [H + 2M + L]/4 \tag{15.14}$$
$$\sigma = [H - L]/2.65, \tag{15.15}$$

where

H = high estimate
L = low estimate
M = most likely value (Mode)
2.65 = value for 80% confidence level.

Note that the expression for the mean for the previous problem would be:

$$\mu = [H + 2M + L]/4 = (20 + 2 \times 12 + 7)/4 = 12.75.$$

This is less than the mean value of 13 for the actual triangular distribution as the mode has a higher weight in the approximation expression. It could be higher or lower if $2M$ is higher or lower than $(H + L)$. Note that the mode has the same weight as the total of the high and low estimates in determining the mean.

The equations used in PERT are very similar to those of C&D, but they assume the high and low values are the actual high and low values with no probability of being outside the limits and a confidence limit of approximately 100%. The standard deviation equation represents the range divided by 6, which is occasionally used to estimate the standard deviation when the normal distribution is used. Those equations for PERT analysis are:

$$\mu = [H + 4M + L]/6 \tag{15.16}$$
$$\sigma = [H - L]/6, \tag{15.17}$$

where

H = high estimate
L = low estimate
M = most likely value (Mode)
6 = value for 100% confidence level.

Note that the expression for the mean for the previous problem would be:

$$\mu = [H + 4M + L]/6 = (20 + 4 \times 12 + 7)/6 = 12.50.$$

This is less than the value of 13 for the actual triangular distribution as the median has a higher weight in the approximation expression. It could be higher or lower if $4M$ is higher or lower than $2 \times (H + L)$. In these formulas the most likely value or mode is used, and if the distribution was normal, the mean and the mode would be equal. Note that the mode has twice the weight of the sum of the high and low estimates in PERT vs. equal weights in the C&D approach.

Triangular Cash Flow Analysis Using Cooper and Davidson Approach

A present worth cash flow analysis, in million dollar units, was performed using the Cooper and Davidson [6] equations using the assumption that there is only a 10% chance that the value will be higher or a 10% chance that it will be lower than the lower limit. This represents a 80% confidence range and the following values were obtained in Table 15.6:

Table 15.6: Cash flow analysis data for C&D approach

	Most Likely	Range (%)	Low Value	High Value	Mean μ	Standard Deviation σ
Revenue	150	−20/+10	120	165	146.25	16.98
Expenses	60	−10/+20	54	72	61.5	6.79
Investment	75	−5/+5	71.25	78.75	75.0	2.83
Cash Flow	+15				μ = +9.75	σ = 18.51

$$\sigma^2 = 16.98^2 + 6.79^2 + 2.83^2 = 342.81$$
$$\sigma = 342.81^{1/2} = 18.51$$
$$P[CF < 0] = \Phi[(0 - 9.75)/18.51] = \Phi[-0.526] \approx 0.30 = 30\%$$
$$MinCFAT = 9.75 - 2.65 \times 18.51/2 = -14.8$$
$$MaxCFAT = 9.75 + 2.65 \times 18.51/2 = +34.3.$$

The range would be −14.8 to 34.3 or 49.1 million dollars.

Traditional analysis indicates that the project will make 15 million dollar units using the most likely values. The C&D risk analysis indicates that, due to the variability in the data, the expected value of the CFAT is only 9.75 million because of the range of cash flow components. Also, there is a 10% chance that the project will have a cash flow of a negative 14.8 million or lower. There is also a 10% chance the project will have a cash flow greater than 34.3 million. The relatively high probability of a negative cash flow at approximately 30% would tend to cause rejection of the project.

Triangular Cash Flow Analysis Using PERT Approach

An analysis using PERT would have a larger range for the higher and lower values as that assumes the lower and upper limits are absolute values; that is there is no chance of higher or lower values. The follow is using estimated values for the higher and lower limits for the 3 standard deviations of the revenues and expenses of 30 and 50% values. These are the upper accuracy range limits for feasibility studies (see Table 15.7).

$$\sigma^2 = 20^2 + 8^2 + 2.5^2 = 470.25$$
$$\sigma = 470.25^{1/2} = 21.69$$

Table 15.7: Cash flow analysis data for PERT approach

	Most Likely	Range (%)	Low Value	High Value	Mean μ	Standard Deviation σ
Revenue	150	-50/+30	75	195	145	20
Expenses	60	-30/+50	42	90	62	8
Investment	75	-10/+10	67.5	82.5	75	2.5
Cash Flow	+15				μ = +8.0	σ = 21.69

$$P[CF < 0] = \Phi[(0 - 8.0/21.69] = \Phi[-0.369] \approx 0.35 = 35\%$$
$$MinCFAT = 8.0 - 6 \times 21.69/2 = -57.1$$
$$MaxCFAT = 8.0 + 6 \times 21.69/2 = +73.1$$

The range would be −57.1 to 73.1 or 130.2 million dollars

Traditional analysis using the most likely values indicates that the project will make the required return plus an additional 15 million dollar units. This PERT risk analysis indicates that, due to the variability in the data, the expected value of the additional CFAT is only 8 million. The estimate range is from a low of −57.1 million to a positive 73.1 million. The range is larger than the C&D analysis as the confidence level would be nearly 100% instead of 80%. The higher probability of a negative cash flow at approximately 35% would tend to cause rejection of the project. The PERT values results in a wider estimate range and higher probability of a negative cash flow. The larger range of the PERT analysis, approximately 130 million dollars, vs. that of the C&D, approximately 50 million dollars, is what had led to the development of the C&D approach.

If the exact same ranges were used for the PERT and C&D models, the PERT would have a smaller range as it divides the range by 6 vs. 2.65 for C&D. The PERT method also puts more weight on the mode in calculating the distribution mean.

15.4   RISK SUMMARY

The basic risk analysis methods in of Sensitivity Analysis and Optimistic-Pessimistic Analysis give a point estimate value, but no indication of the range of possible error. The sensitivity analysis gives an indication of the effect of critical variables upon the engineering economic expression by examining specific cases to determine which variables are critical for cost control of the process. The optimistic-pessimistic analysis is generally restricted to two variables to determine which variable is more important.

The discrete probability analysis requires that the discrete probability density function for each of the possible events be known and when the number of events is numerous, it can be tedious to evaluate. However, the mean can be calculated to determine the expected profit of the process and the probability of a loss can be determined by summing the probabilities of those events in which a loss occurs.

The continuous probability analysis uses the data and the probability density function for that data set to calculate the mean and standard deviation. The probability of an individual item has zero probability, so an event represents a continuous range of events. The two primary density functions for the analysis of cost data are the triangular distribution and the normal distribution. The normal distribution can be used for analyzing the sum of individual distributions which may not be normal, but will approach the normal distributions via the central limit theorem. The triangular distribution is better for estimating individual revenue and cost components as the most likely value is usually not the mean in cost analysis and the total cost can be estimated with the normal distribution via the central limit theorem.

The advantage of probability analysis is that it is able not only to determine a range but also determine the probability of a loss. The PERT method is applied more and has a higher confidence level than the C&D approach but the longer range of PERT tends to lower the estimated mean and results in a higher standard deviation leading to a higher probability of a loss. The variability of the estimate and confidence level of the estimate are important in evaluating the risk of the project.

15.5   REFERENCES

[1] Lindgren, B. W. and McElrath, G. W., *Introduction to Probability and Statistics*, The McMillan Company, New York, p. 277, 1959. 227

[2] Garvey, Paul R., *Probability Methods for Cost Uncertainty Analysis*, Marcel Dekker, Inc, New York, pp. 109–112, 2000. 227, 235

[3] Garvey, Paul R., *Probability Methods for Cost Uncertainty Analysis*, Marcel Dekker, Inc, New York, pp. 51–53, 2000. 229

[4] Wikipedia Web Page, (2-17-2018). https://en.wikipedia.org/wiki/Triangular_distribution 235

[5] Wikipedia Web Page, (2-17-2018). https://en.wikipedia.org/wiki/Program_evaluation_and_review_technique 236

[6] Cooper, D. O. and Davidson, L. B., The parameter method for risk analysis, *Chemical Engineering Progress*, pp. 73–78, November 1976. 236, 238

15.6   EVALUATIVE QUESTIONS

1. Use the following discrete data matrix, complete the matrix (Table 15.8) and calculate the mean and standard deviation of the Present Worth (15%).

$$PW(15\%) = 2{,}457{,}204 \times SP(\$/kg) - 85{,}859{,}244$$

Table 15.8: Selling price data matrix

Sales Price $/Kg	Probability	Present Worth (15%)	Expected Selling Price 5/Kg	E(x²) for Variance	Expected Present Worth (15%)	E(x²) for Variance of PW (15%)
x	p(x)	PW/10³	x*p(x)	x² * p(x)		
30	0.10					
34	0.30					
38	0.30					
42	0.30					
46	0.10					
Totals	1.00					

μ (Selling Price) =_____   μ (Present Worth (15%)) =_____
σ² (Selling Price) = _____   σ² (Present Worth (15%)) = _____

2. Donnie the Dealmaker has a revised set of revenue probabilities. Complete Table 15.9 and answer the following questions.

   (a) What is the expected profit?.

   (b) What is the variance of the profit?

   (c) What is the standard deviation of the profit?

   (d) What is the range for the possible profit scenarios?

   (e) What is the actual probability of a loss?

   (f) What is the probability of a loss using the normal distribution?

3. The mean of a normal distribution is 1,000 and the standard deviation is 100.

   (a) What is the probability that a random variable selected from that distribution is less than 700?

   (b) What is the probability that the random variable selected is less than 900?

Table 15.9: Revenues, costs, and profits for Donnie the Dealmaker's revised product investment

Revenue Amount	Revenue Probability	Cost Amount	Cost Probability	Profit Amount	Profit Probability	Expected Profit	E(X²)
40	0.3	30	0.2	10	0.06	0.6	6
40	0.3	40	0.5	0	0.15	0.0	0
40	0.3	50	0.3	-10	0.09	-0.9	9
50	0.5	30	0.2	20			
50	0.5	40	0.3	10			
50	0.5	50	0.5	0			
60	0.2	30	0.2	30			
60	0.2	40	0.3	20			
60	0.2	50	0.5	10			
Totals					1.00		

(c) What is the probability that the random variable selected is greater than or equal to 1,200?

(d) What is the probability that the random variable is between 900 and 1,100?

4. A variable follows the triangular distribution and has the lower limit of 600, an upper limit of 2,200 and the most likely value of 1,200.

(a) What is the mean?

(b) What is the standard deviation?

(c) What is the probability that the variable is less than 800?

(d) What is the probability that the variable is greater than 1,500?

(e) What is the probability that the variable is between 800 and 1,500?

5. The cash flows for a project with an initial investment of $20,000 are in Table 15.10.

(a) If the MARR is 10%, what is the present worth of the cash flows?

(b) What is the variance of the cash flows?

(c) What is the standard deviation of the cash flows?

(d) What is the probability that the cash flows will be negative?

(e) What is the probability that the cash flow is less than 2000?

Table 15.10: Cash flows for a project with an initial investment of $20,000

Year	Expected Cash Flow μ	Standard Deviation of Cash Flow σ	Variance of Cash Flow σ^2
0	-20,000	100	10,000
1	8,000	200	40.000
2	9,000	300	90,000
3	10,000	400	160,000

6. The values in Table 15.11 were obtained is the response to the bids for a contract. The bids are assumed to follow a normal distribution.

   (a) What is the actual range of the bids?

   (b) What is the mean value of the bid distribution?

   (c) What is the standard deviation for the bid distribution?

   (d) What is the probability that a bid would be less than $350,000?

   (e) What is the probability that a bid would be greater than $450,000?

Table 15.11: Bids for a contract

Bid Number	Estimate (1,000 $ Units)
0	350
1	420
2	375
3	390
5	370

7. (a) Cost/value engineers were sent back to re-examine a project with the new data provided. (See Table 15.12.) Using the C&D approach, determine the "risk" and would you recommend the project?

   – What is the mean value?

   – What is the standard deviation?

   – What is the probability the cash flow is negative?

- What is the probability that the cash flow is less than 10?
- What is the probability the cash flow is less than 20?
- Why or why not do you recommend the project?

Table 15.12: Cash flow analysis data for C&D approach

	Most Likely	Range (%)	Low Value	High Value	Mean μ	Standard Deviation σ
Revenue	140	–15/+5				
Expenses	50	–10/+15				
Investment	70	–5/+5				
Cash Flow	+20				μ =	σ =

(b) Using the PERT approach (see Table 15.13), determine the "risk" and would you recommend the project?

- What is the mean value?
- What is the standard deviation?
- What is the probability the cash flow is negative?
- What is the probability that the cash flow is less than 10?
- What is the probability the cash flow is less than 20?
- Why or why not do you recommend the project

Table 15.13: Cash flow analysis data for PERT approach

	Most Likely	Range (%)	Low Value	High Value	Mean μ	Standard Deviation σ
Revenue	140	–15/+5				
Expenses	50	–10/+15				
Investment	70	–5/+5				
Cash Flow	+20				μ =	σ =

8. The cash flows from a project are presented in Table 15.14. The required rate of return is 10%. What is the present worth of the expected cash flow on the project and what is the probability that the cash flow from the project has a loss?

(a) What is the present worth of the expected cash flows at a MARR of 10%?

(b) What is the standard deviation of the cash flows for the project?

(c) What is the probability the project will have zero cash flows with the MARR of 10%?

(d) What is the probability the project will have zero cash flows with the MARR at 20%?

Table 15.14: Cash flows from a project

Year	Expected Cash Flow μ($)	Standard Deviation of Cash Flow μ	Variance of Cash Flow μ^2
0	-9,000	100	10,000
1	4,000	200	40,000
2	4,000	400	160,000
3	4,000	600	360,000

APPENDIX A

Discrete and Continuous Compounding Factors

Please see the tables on the pages that follow.

Table A.1: Discrete compounding factors of economic expressions—Discrete payments and discrete interest

Payment Type	Factor Name	Find	Given	Symbol	Formula
A. Single Payment	Present Worth	P	F	(P/F, i, n)	$(1+i)^{-n}$
	Future Worth (Compound Amount)	F	P	(F/P, i, n)	$(1+i)^n$
B. Uniform Payment (Uniform Series)	Sinking Fund	A	F	(A/F, i, n)	$i/[(1+i)^n -1]$
	Capital Recovery	A	P	(A/P, i, n)	$[(i(1+i)^n)]/[(1+i)^n -1]$
	Compound Amount	F	A	(F/A, i, n)	$[(1+i)^n -1]/i$
	Present Worth	P	A	(P/A, i, n)	$[(1+i)^n -1]/[i(1+i)^n]$
C. Uniform Gradient Expression					
Standard Uniform Gradient	Uniform Gradient Present Worth	P	G	(P/G, i, n)	$[((1+i)^n -1 -ni)/(i^2 (1+i)^n)]$
	Uniform Gradient Future Worth	F	G	(F/G, i, n)	$[((1+i)^n -1 -ni)/i^2]$
	Uniform Gradient Uniform Series	A	G	(A/G, i, n)	$[((1+i)^n -1 -ni)/((1+i)^n -1)]$
Uniform Ramp Gradient	Uniform Ramp Gradient Present Worth	P_R	G	$(P_R/G, i, n)$	$[((1+i)^{n+1} -1 -i(n+1))/(i^2 (1+i)^n)]$
	Uniform Ramp Gradient Future Worth	F_R	G	$(F_R/G, i, n)$	$[((1+i)^{n+1} -1 -i(n+1))/(i^2)]$
	Uniform Ramp Gradient Uniform Series	A_R	G	$(A_R/G, i, n)$	$[((1+i)^{n+1} -1 -i(n+1))/(i(1+i)^n -1)]$
D. Geometric Gradient Expression					
Geometric Gradient	Geometric Gradient Present Worth	P	A_1, g	$(P/A_1 g, i, n)$	$[(1-((1+g)^n/(1+i)^n))/(i-g)]$
			If g=i	$(P/A_1, g=i, n)$	$n/(1+i)$
	Geometric Gradient Future Worth	F	A_1, g	$(F/A_1, g, i, n)$	$[((1+i)^n - (1+g)^n)]/[i-g]$
			If g=i	$(F/A_1, g=i, n)$	$n(1+i)^{(n-1)}$
	Geometric Gradient Uniform Series	A	A_1, g	$(A/A_1, g, i, n)$	$[(i((1+i)^n -(1+g)^n))/((i-g)((1+i)^n -1))]$
			If g=i	$(A/A_1, g=i, n)$	$[ni(1+i)^{(n-1)}]/[(1+i)^n -1]$
Escalation Gradient	Escalation Gradient Present Worth	P_E	A_1, E	$(P_E/A_1, E, i, n)$	$[(1+E)/(E-i)][((1+E)/(1+i))^n -1]$
			If E=i	$(P_E/A_1, E=i, n)$	n
	Escalation Gradient Future Worth	F_E	A_1, E	$(F_E/A_1, E, i, n)$	$[(1+E)/(E-i)][((1+E)^n - (1+i)^n]$
			If E=i	$(F_E/A_1, E=i, n)$	$n(1+i)^n$
	Escalation Gradient Uniform Series	A_E	A_1, E	$(A_E/A_1, E, i, n)$	$[(i(1+E)/(E-i))*((1+E)^n - (1+i)^n)]/[(1+i)^n -1]$
			If E=i	$(A_E/A_1, E=i, n)$	$ni(1+i)^n/[(1+i)^n -1]$

Notation:
P=Present Worth;   i = effective discrete interest rate per period;   A=uniform end-of-period payments;   n = number of periods;
F=Future Worth;   g=Geometric Gradient Rate;   G=Uniform Gradient Amount;   E = Escalation Gradient Rate;
A_1 = Initial Geometric Gradient Amount and Initial Escalation Gradient Amount

appendix1

Table A.2: Continuous compounding factors of economic expressions—Discrete payments and continuous interest

Payment Type	Factor Name	Find	Given	Symbol	Formula
A. Single Payment					
	Present Worth	P	F	(P/F,r,n)	e^{-rn}
	Future Worth	F	P	(F/P,r,n)	e^{rn}
B. Uniform Payment (Uniform Series)					
	Sinking Fund	A	F	(A/F,r,n)	$[(e^r-1)/(e^{rn}-1)]$
	Capital Recovery	A	P	(A/P,r,n)	$[e^{rn}(e^r-1)/(e^{rn}-1)]$
	Future Worth	F	A	(F/A,r,n)	$[(e^{rn}-1)/(e^r-1)]$
	Present Worth	P	A	(P/A,r,n)	$[(e^{rn}-1)/(e^{rn}(e^r-1))]$
C. Uniform Gradient Expressions					
Standard Uniform Gradient	Uniform Gradient Present Worth	P	G	(P/G,r,n)	$\{[(e^{rn}-1)-n(e^r-1)]/[(e^r-1)^2\,e^{rn}]\}$
	Uniform Gradient Future Worth	F	G	(F/G,r,n)	$\{[(e^{rn}-1)-n(e^r-1)]/[(e^r-1)^2]\}$
	Uniform Gradient Uniform Series	A	G	(A/G,r,n)	$\{[(e^{rn}-1)-n(e^r-1)]/[(e^r-1)(e^{rn}-1)]\}$
Uniform Ramp Gradient	Uniform Ramp Gradient Present Worth	P_R	G	$(P_R/G,r,n)$	$\{[(e^{r((n+1)}-1)-(n+1)(e^r-1)]/[(e^r-1)^2(e^{rn})]\}$
	Uniform Ramp Gradient Future Worth	F_R	G	$(F_R/G,r,n)$	$\{[(e^{r((n+1)}-1)-(n+1)(e^r-1)]/[(e^r-1)^2]\}$
	Uniform Ramp Gradient Uniform Series	A_R	G	$(A_R/G,r,n)$	$\{[(e^{((n+1)}-1)-(n+1)(e^r-1)]/[(e^r-1)^2(e^{rn-1})]\}$
D. Geometric Gradient Expressions					
Geometric Gradient	Geometric Gradient Present Worth	P	A_1,b	$(P/A_1,b,r,n)$	$\{[1-(e^{bn}/e^{rn})]/[e^r-e^b]\}$
			If b=r	$(P/A_1,b=r,n)$	n/e^r
	Geometric Gradient Future Worth	F	A_1,b	$(F/A_1,b,r,n)$	$\{[e^{rn}-e^{bn}]/[e^r-e^b]\}$
			If b=r	$(F/A_1,b=r,n)$	$n/e^{r(n-1)}$
	Geometric Gradient Uniform Series	A	A_1,b	$(A/A_1,b,r,n)$	$\{[e^{rn}-e^{bn}]/[e^r-e^b]\}\,[[e^r-1)/(e^{rn}-1)]]$
			If b=r	$(A/A_1,b=r,n)$	$[n[(e^{rn})/(e^{rn}-1)]*[(e^r-1)/(e^r)]$
Escalation Gradient	Escalation Gradient Present Worth	P_E	A_1,c	$(P_E/A_1,c,r,n)$	$\{[((e^c)/(e^c-e^r)]*[(e^{cn}-e^{rn})e^{rn}]$
			If c=r	$(P_E/A_1,c=r,n)$	n
	Escalation Gradient Future Worth	F_E	A_1,c	$(F_E/A_1,c,r,n)$	$\{[((e^c)/(e^c-e^r)]*[(e^{cn}-e^{rn})]$
			If c=r	$(F_E/A_1,c=r,n)$	ne^{rn}
	Escalation Gradient Uniform Series	A_E	A_1,c	$(A_E/A_1,c,r,n)$	$\{[((e^r-1)(e^c)/(e^c-e^r)*[(e^{cn}-e^{rn})/(e^{rn}-1)]\}$
			If c=r	$(A_E/A_1,c=r,n)$	$\{[n(e^r-1)e^{rn}/(e^{rn}-1)\}$

Notation:
P=Present Worth;   i = effective discrete interest rate per period;   A=uniform end-of-period payments;   n = number of periods;
F=Future Worth;   g=Geometric Gradient Rate;   G=Uniform Gradient Amount;   ε = Escalation Gradient Rate;
A_1 = Initial Geometric Gradient Amount and Initial Escalation Gradient Amount

Author's Biography

ROBERT C. CREESE

Dr. Robert C. Creese was Professor of Industrial and Management Systems Engineering at West Virginia University and taught courses on Engineering Economy, Advanced Engineering Economics, Cost and Estimating for Manufacturing, Manufacturing Processes, and Advanced Manufacturing Processes. He has previously taught at The Pennsylvania State University (9 years), Grove City College (4 years), Aalborg University in Denmark (3 sabbaticals), and West Virginia University for 35 years. He worked at US Steel for two years as an Industrial Engineer before starting his teaching career.

Dr. Creese is a Fellow of the AACE International, received the Charles V. Keane Service Award and Brian D. Dunfield Educational Service Award presented by AACE International, and was treasurer of the Northern West Virginia Section of AACE International for more than 20 years. He is a Life Member of AACE International, ASEE (American Society for Engineering Education), and ASM (American Society for Materials). He also is a member of ICEAA (International Cost Estimating & Analysis Association), AIST (Association for Iron & Steel Technology), AWS (American Welding Society), and AFS (American Foundry Society).

He obtained his B.S. in Industrial Engineering from The Pennsylvania State University, his M.S. in Industrial Engineering from the University of California at Berkeley, and his Ph.D. in Metallurgy from The Pennsylvania State University.

Dr. Robert C. Creese has authored the book *Introduction to Manufacturing Processes and Materials* (Marcel Dekker-1999) and co-authored two books *Estimating and Costing for the Metal Manufacturing Industries* (Marcel Dekker-1992) with Dr. M. Adithan, Professor Emeritus of VIT University Vellore, India, and Dr. B.S. Pabla of the Technical Teachers' Training Institute, Chandigarh, India, and *Strategic Cost Analysis for Project Managers and Engineers* (New Age International Publishers-2010) with Dr. M. Adithan, VIT University, Vellore, India. He has authored/co-authored more than 100 technical papers.

Printed in the United States
by Baker & Taylor Publisher Services